THE FORD-WYOMING DRIVE-IN

CARS, CANDY & CANOODLING IN THE MOTOR CITY

KAREN DYBIS

Charleston • London

THE
History
PRESS

Published by The History Press
Charleston, SC 29403
www.historypress.net

Copyright © 2014 by Karen Dybis
All rights reserved

First published 2014

ISBN 978.1.5402.1048.7

Library of Congress CIP data applied for.

To the Clark and Shafer families, my family and every other family that falls in love with the Ford-Wyoming.

CONTENTS

ACKNOWLEDGEMENTS

I have written you down.
Now you will live forever.

—Bastille, "Poet"

This book came about because I wrote an article for the *Detroit News* on the issue of digital conversion at drive-in theaters. Knowing the Ford-Wyoming was the largest around, I called on Charles Shafer for the story. Our conversations were a delight. That made it easy to agree to this project. Who knew that there was so much more to say and write? Thank you to Greg Tasker at the *News* for such a life-changing assignment.

It has been my honor to have worked with Shafer, a magnetic and accomplished movie-loving man. Many thanks go as well to Bill Clark and Virgil Berean, two of the finest gentlemen around. You all were extremely gracious in sharing your time and stories with me.

Few words can express the honor and privilege it was to get to know the Clark family. Although James, Clyde Jr. and Harold are gone, they can be extremely proud of how their children, grandchildren and great-grandchildren hold the Ford-Wyoming dear. The Clarks' contribution to this story was monumental. A special thanks goes to Diane Clark O'Brien, who is the heart of this book. Her work finding photos, sources and anecdotes brought the story alive.

To Karen Wisniewski, David L. Good and the staff at the Dearborn Historical Museum. You were the first people I turned to for information and guidance. Your help was huge in finding the Clark family, Orville L. Hubbard and the high-flying truth about Haggerty Airfield.

To Richard Story and the Wayne Historical Society. Finding your treasure-trove of information about Charles and Martin Shafer was like hitting the jackpot. Your resources and institutional knowledge of the Shafer family was a game-changer.

To Carol Delves of the Howe-Peterson Funeral Home. Your kindness in responding to a crazy e-mail request from a frustrated writer can never be truly repaid. If not for you, the Clark family's story might not have been told. And thanks to Doug Clark for responding.

To Pamela Myers-Grewell. Your incredibly detailed and helpful information on Sligo, Pennsylvania; its history; and Clyde W. Clark Sr.'s genealogy is amazing.

To my family. To my parents, Duane and Carolyn Talaski, for inspiring my love of reading and supporting my desire to be a writer. To Mark, Pete and Robin, who took this long walk with me. You are loved beyond compare.

A FAREWELL TOUR

It was a clear June morning when the procession carrying Harold Clark made its final inspection of the Ford-Wyoming Drive-in Theatre.

The black hearse had been past each significant landmark in the lifelong Dearborn resident's life: his childhood home on Steadman Street, the Town-N-Country Bowling Lanes in nearby Westland and the former site of the Dearborn Tool & Machine on Ford Road. One landmark remained on the schedule: the iconic "ozoner," or drive-in theater, by the corner of Ford Road and Wyoming Street that Clark and his brothers built.

Clark's travels to the Ford-Wyoming that day in 2012 also took him past the statue of former Dearborn mayor Orville L. Hubbard. His figure still looms large in front of city hall on Michigan Avenue. Hubbard—that bombastic, outspoken politician and perfectionist—was the most vocal opponent to the drive-in's construction. Hubbard's many attempts to stymie the Ford-Wyoming put the Clark brothers at odds with the powerful mayor at a time when few others dared to challenge him.

Steven, Clark's youngest son, had asked for permission to take his father through the Ford-Wyoming that day. Although the Clark family had sold the drive-in nearly thirty years before, it remained part of their father's legacy. This would be his last goodbye.

The question didn't need to be asked. Of course, "Uncle Charlie" said. Clark's longtime friend and current Ford-Wyoming owner Charles Shafer was also mourning the passing of his late-night dinner companion and business confidant. It was the least he could do, Shafer said, for the family who had built the theater and been its guardian for three decades.

No one documented that cruise through the Ford-Wyoming, the hearse's wheels treading heavy on the clay and stone. The property once served as the foundation for one of Dearborn's preeminent brickyards and, later, an amateur airfield. The neighboring businesses likely chugged through their daily routines, never noticing the lineup of cars in the driveway. After all, vehicles waiting for their turn to go inside were part and parcel of the Ford-Wyoming, even after all of these years.

The theater's manager, Virgil Berean, would later ask about the procession, wondering if the rumored drive actually took place. Berean wasn't there to watch Clark make his final turn onto Ford Road. Much like those men who came before him, he was working that night, arriving just as the sun was falling and staying until nearly daylight. During that long night shift, Berean would check the screens, reattach speakers, pick up litter and prepare for another crowd. In a very real way, his work as the Ford-Wyoming's caretaker is a kind of tribute to Clark and Shafer for their devotion to this institution.

Although Clark didn't request it, it is understandable why his family wanted that final farewell. Something about this drive-in theater stays with you, whether it is its mighty height, its dusty acres or its oddly peaceful location off I-94 on the border of Detroit and Dearborn. There is a magic there, an alchemy that combines the new with the old. It is the memories of children running to the playground under the original screen's shadow. It is the smell of popcorn, hotdogs and candy bars. It is the glare of neon signs advising drivers, "Children Playing Drive Slow." Despite its rust that comes with six decades of hard work, there is life.

Maybe it is nostalgia or an appreciation for kitsch, but every generation seems to find the Ford-Wyoming. It is Metro Detroit's community-gathering spot. It knows no race, color or creed. It serves only to entertain. As Shafer promises, it will continue to do so for years to come.

CHAPTER 1

A FIELD OF DESTINY

Wanna know all the things that I shouldn't know
I got smokes that I stole from the Seven-L
Everybody's at the drive-in, wanna go?
—*Ray LaMontagne, "Drive-in Movies"*

The story of the Ford-Wyoming Drive-in starts in a simple way—it begins with the land. To the naked eye, it looks like the surface of the moon. It is dusty and pitted. Pipes buried deep within it are always leaking, leaving muddy trails in their wake. Brick shards turn up frequently. It is, for the most part, inhospitable to anything except a drive-in theater.

There is good reason for that. For much of its past, men have worked it, hard. They tilled it for crops. They dug craters into it. They moved machines across it. They landed planes on it. It sits among truck yards and gasoline tanks, an industrial site in a city known for manufacturing.

Long before John Wayne graced the Ford-Wyoming's tower and teenagers necked below it, the lot at Ford Road and Wyoming Street served as one of the many family farms that populated the area. Later, it was part of the Haggerty Brick Company, pumping out enough building material to support Dearborn and neighboring Detroit for decades.

Once the land was mined of its natural resources, brick magnate John S. Haggerty loaned it out to serve as an airfield, providing a landing site for the likes of World War I flying ace Eddie Rickenbacker. Eventually, it would end up in the hands of Clyde W. Clark Sr. and his three sons to build a drive-in theater.

But we'll get to that soon enough.

Let's start with a little about Dearborn. It is best known as hometown of automotive innovator Henry Ford, whose company is headquartered there. The legendary industrialist moved from Detroit to an area that was then known as Springwells Township to build a factory near the Rouge River. With his five-dollar-a-day wage offering, Ford would not only draw all of Metro Detroit to the area but also bring the world to his small hometown, expanding its population tenfold.

Wait. Let's go back even farther. Before there was Ford, there were farmers. They, too, chose to live along the Rouge toward the late 1700s. The Dumais, Drouillard and Cissne families are said to have set up what is now modern-day Dearborn in 1786.

The city's name came from Lewis Cass, who was governor of what was then known as the Michigan Territory. Cass christened the area after his close friend, Major General Henry Dearborn, a New Hampshire physician who served as secretary of war under President Thomas Jefferson.

The area grew quickly during the nineteenth century. Two items of note: In 1825, construction began on what was known as Chicago Road, a military territorial highway that would become Michigan Avenue. A few years later, Detroit moved its arsenal out of the city and onto Michigan Avenue. The Detroit Arsenal's main purpose was to serve as a supply depot for the army; it stored, maintained and repaired arms and ammunition.

With the population that followed, the town of Dearborn gained enough residents to incorporate in 1893. Like most settlers, Dearborn's earliest industrialists used what they could find around them to build new businesses. Rich forests resulted in a booming timber industry. Textiles were another significant contributor, especially Arna Mills, which produced women's clothing.

And then there was clay. Lots and lots of clay. Developments such as the massive Detroit Arsenal needed materials for construction, and brickyards sprung up to get their share of the business. Brickyards, it could be argued, were Dearborn's first manufacturing industry, long before that Ford character started to make quadcycles and fancy automobiles. Brickyards began popping up toward the end of the nineteenth century. Titus Dort, a Vermont native, came to Dearborn in 1829 and his brickyard is recognized as the first mostly because of his significant production and years in business.

But back to that Haggerty character and the Ford-Wyoming. John Strong Haggerty was born on August 22, 1866. He was the son of Lorenzo Dow Haggerty and Elizabeth (Strong) Haggerty, both Michigan natives. His

Artist's rendering of the Haggerty Brickyards and kilns in Dearborn. *Dearborn Historical Museum.*

Workers display the bricks in production at the Haggerty Brickyard. The photo is signed by John Strong Haggerty, the brickyard's owner. *Dearborn Historical Museum.*

grandfather Hugh Henry Haggerty came to Michigan from Ireland by way of New York. Hugh Haggerty was one of the area's first settlers when he moved there in 1828. He purchased land from the government and set up what was described as a "productive farm."

John's father is considered one of the "pioneer brick makers" of his time. He cofounded the Haggerty & Proctor brickyard on a site near Michigan

Avenue and Central Street with his nephew, William Proctor, making bricks by hand and setting them in the sun to dry. Demand was high because of the area's growing population, so the business thrived.

In those early years, brickyard workers would remove the top soil to uncover the clay beneath. Workers dug the clay out by hand, going about eight feet deep to start. They then loaded the clay into horse-drawn carts. These cartloads were dumped into pits and soaked with water overnight. The next day, the clay was scooped into an elevator, which carried it to a brick machine. Its interior knives would cut and mix the clay. This clay would be pressed into oblong molds, which were then loaded onto hand trucks and carted to the drying yards. There, the bricks were sun-dried, a process that usually took at least a day. Because the sun did most of the work, brickmaking became a seasonal business. Afterward, the bricks were taken to a kiln and baked over wood fires.

John was about twenty years old when he followed in his father's footsteps and started his first brickyard in the spring of 1887. By then, there were nineteen manufacturers in the area, and Dearborn became known as the largest producer of common brick for the state of Michigan.

Haggerty started the business with his twin brother, Clifton. Haggerty's first brickyard was built on the site of the longtime family farm, which was on Michigan Avenue west of Wyoming Street. Years later, he started another yard on Ford Road east of Miller Road. The plant at 10600 Ford Road was an estimated fifty-five acres.

Haggerty became known for using more "modern" techniques. Haggerty brickyards used steam shovels and engine-operated gauge cars to cart the clay away. He constructed buildings, taking the impact of the weather and elements out of the process. Estimates say that Haggerty's yards produced about 1.5 million bricks in their first year. At their peak, the Haggerty brickyards employed about 150 people and had an annual production of between 40 to 60 million bricks.

The Great Depression hit all businesses hard, the brickyards especially so. The 1920s boom was over, and the construction business was largely silenced. Haggerty brick was used in some New Deal project works in Wayne County, probably because Haggerty himself was such a powerful character in that county's political life. Not only did he serve as a Wayne County road commissioner, Haggerty was also at one time president of the Michigan State Fair Association, a Michigan delegate to the Republican National Convention, Michigan Republican State Central Committee member, Michigan Republican Party treasurer (1927–29) and Michigan's secretary

JOHN S. HAGGERTY
DETROIT
L. D. HAGGERTY & SON
BRICK MANUFACTURERS

John Strong Haggerty was considered one of the largest brick manufacturers in Dearborn, owning two brickyards there. *Dearborn Historical Museum.*

The property at Ford Road and Wyoming Street was largely undeveloped after brick production slowed. It served as an airfield throughout World War II. *Dearborn Historical Museum.*

of state (1927–30). He also socialized with the city's elite as a member of the Detroit Yacht Club, Detroit Athletic Club and others.

However useful they were to the community, brickyards were environmentally devastating. The land was left with gaping pits and shallow ponds once the clay was removed. Brickyards often filled them with dirt obtained from city excavations and private businesses. The ground was leveled and left as open fields.

In the early 1920s, Edward "Eddie" Rickenbacker had an idea of what to do with some of that land. According to airport historian Dick Soules, Rickenbacker came into town frequently as he had a partnership to build his namesake automobile. Rickenbacker at the time was both a world-famous race car driver and a World War I flying ace, earning him the Congressional Medal of Honor.

Rickenbacker wanted a place to land nearby so he could come and go for meetings. He talked to Haggerty, who allowed him to turn the northeast side of the brickyard into an airstrip. The location was ideal: the automotive plant where Rickenbacker automobiles were produced was on Cabot Street in Detroit, just a few miles away from the brickyard site.

The Rickenbacker Motor Company was incorporated in July 1921 as a partnership between Rickenbacker and auto veterans, including Bryon F. "Barney" Everitt, a custom auto-body builder from Detroit. Rickenbacker became the vice-president and director of sales. Despite its potential, the company only lasted five short years. Internal fighting and production issues caused Rickenbacker to resign in September 1926. Rickenbacker Motor closed in February 1927. The company produced an estimated 34,500 cars; about 100 are believed to still exist today.

Although Rickenbacker moved on, the eighty-acre airfield on Wyoming remained popular among amateur pilots. One in particular was Leonard Flo, president of Flo Flying Service, who leased the field from Haggerty. Flo and fellow pilots used the field frequently, even hosting a flying club there called the Wise Birds. He would offer rides for fifty cents, Soules recalled. Flo even had a nearby flying school, Leonard Flo Air College, which frequently included Rickenbacker in its advertisements.

By now, the brickyards and most of their equipment were largely gone. Gasoline company Mobil bought a chunk of the land for a tank farm. Henry Ford—Haggerty's childhood friend from their days as schoolmates at Springwells District School No. 5—would move clay, bricks and several pieces of equipment to Greenfield Village, a historical museum that he was constructing in Dearborn.

An estimated fifty truckloads of Haggerty clay was delivered to the Village to create bricks on site at the Haggerty Power Plant, which was dedicated in 1937. These bricks were used throughout the Village, allowing visitors to watch its day-to-day construction.

The power plant was retired in the 1960s to become a pottery shop with on-site demonstrations. As the Village's layout changed over time, the pottery shop was located too far off the main areas, so it moved to a more central location, according to Melinda Mercer, who has been a full-time potter at Greenfield Village since 2004. The Haggery facility is now a "Shipping and Receiving" depot, hidden behind the Firestone Farm in a nonpublic area. But Haggerty's signature, forever written in concrete, is still visible above the building's front door. To this day, every piece of pottery made in Greenfield Village is labeled with an "H" in honor of the Haggerty history there, Mercer said.

At other times, the airfield was used to house special events that passed through Dearborn, such as rodeos, carnivals and even an amusement park, recalled Connie Clark Norsworthy, James A. Clark's daughter.

That large, open area that served as Haggerty Airport would have to bow to progress. Detroit and Dearborn were adding population and new companies quickly. The empty land at Ford and Wyoming attracted the eye of another business, Dearborn Tool & Die. Owner Clyde W. Clark Sr. had moved his successful company to a neighboring site, growing his business along with his family.

After World War II, Clark and his three sons were itching to start something new. Given people's interest in movies, the average household's disposable income, a coming baby boom and a burgeoning car culture in Metro Detroit, they saw opportunity in such a huge property.

Opposite, top: The Ford-Wyoming Drive-in was built near an automotive manufacturing plant and the Dearborn Tool & Die Company at 10400 Ford Road. *Dearborn Historical Museum.*

Opposite, bottom: The drive-in theater (left) was located outside the city's residential areas within an industrial area, but it was convenient to freeways. *Dearborn Historical Museum.*

CHAPTER 2

THE FOUNDING FAMILY

Old days, good times I remember
Fun days filled with simple pleasures
Drive-in movies, comic books and blue jeans
Howdy Doody, baseball cards and birthdays.
—Chicago, "Old Days"

The Clark family was known for many things in and around Dearborn. They were business owners, philanthropists and community boosters. Most of all, they were men of ideas and action. And beware anyone who came between them and their goal.

Clyde Walter Clark Sr. was the family patriarch. He was known in the automotive industry, having founded the Dearborn Tool & Die Company in 1932. There, his family said that he channeled a sixth-grade education into a thriving enterprise, earning several U.S. patents for his inventions. His tools continue to be used today, particularly a hardness tester that he patented in 1943 that determined the hardness of steel for structural integrity.

Born on April 20, 1891, Clark Sr. was raised in Sligo, Pennsylvania, a small coal mining town that was named after the town and county of Sligo in Ireland. The Pennsylvania town's population at the time was a few shy of five hundred. Clark Sr.'s father, Alvin, was a coal miner, as was his father, Seth.

Diane Clark O'Brien, Clark Sr.'s granddaughter, said that the family struck by tragedy when his mother and baby sister died in a house fire while he was at school. This tragedy forced Clark Sr. to grow up quickly, as his father was injured in the fire as he tried to save his wife, according to Connie Clark Norsworthy, his oldest granddaughter. Because his father was in the hospital for months, Clark Sr. had to leave school early on and take odd jobs to keep the family going until his father came home.

That rough start could be why Clark Sr. was so committed to giving his family a peaceful and well-appointed childhood, Connie noted. He also was self-taught in math and other subjects, which explains why her grandfather believed so strongly in higher education, she said.

Clark Sr. tried his hand at the Sligo mines, but it was hard, demanding work. With his inquisitive mind, he was more interested in railroads. He was working in that industry when he met Marie Pearl Crotty on a trip through Butte, Montana. The courtship was swift; they eloped and lived in Idaho and later in Ohio, according to Connie. Marie also had a harsh childhood growing up with her grandmother, Connie said, so the couple had similar backgrounds and beliefs. Much like her husband, Marie was loving to her children, as well as strict and uncompromising. "One always knew where they stood with my grandmother," James A. "Jamie" Clark recalled. "She would not spare words. But she never used profanity; she didn't have to."

Clark Sr. felt the pull of the burgeoning automotive industry, so the family moved again, ending up in Michigan. O'Brien said that her grandfather taught mathematics for a short time at the Henry Ford Trade School, a high school for "needy boys" that operated from 1916 through 1952, graduating more than eight thousand tradesmen before its closure.

Clark Sr.'s skillful tinkering led him to leave this position and start the Dearborn Tool & Die Company. Its original location was at Michigan Avenue and Schafer, next door to what would become Montgomery Ward on the southeast corner, said his grandson Doug Clark. The company would move to its permanent site at 10200 Ford Road soon after, giving Clark Sr. the room he wanted to build a business and pursue his other interests.

Dearborn Tool & Die was known for designing and engineering prototypes and components for heavy machines, gauges and fixtures. It had in-house facilities for jig-grinding, milling, boring and turning, as well as crane capabilities to accommodate up to fifteen tons. "Everything needed to do the job is under one roof," one of its advertisements raved. "Serving industry with fine machines," noted another. It eventually was renamed Dearborn Tool & Machine when Clark Sr.'s sons were running it full time.

His three sons—James, Clyde and Harold—were all born singers and musicians. They also had the same inquisitive nature as their father. James A. Clark was born in 1914. Clyde W. Clark Jr. followed in 1917, and Harold

Clyde W. Clark Sr. took a sixth-grade education and turned it into a thriving business called Dearborn Tool & Die. *Clark family collection.*

Seth Clark arrived in 1925. They lived in a small bungalow in Dearborn on Stedman Street for much of their childhood, which was full of golfing, camping, trips to the beach and other boy-friendly adventures. O'Brien remembers her father, Harold, talking about his first airplane ride at the Haggerty Airfield next door to the Dearborn Tool & Die.

Beyond work, Clyde Sr.'s true passion was his horses. Throughout family photographs, the sometimes stern look on his face melted and a relaxed smile replaced it whenever he had one of his trotters in hand. The land on Ford Road allowed him to keep a few horses and a small racetrack on the property behind the tool shop, said Jamie Clark, James Clark's son.

Around the mid-1940s, Clark Sr. moved to the Ann Arbor/Ypsilanti area in part to pursue his love of horses and harness racing. He purchased a five-bedroom Colonial house at 6200 Geddes Road that Henry Ford had built for former University of Michigan football coach Harry Kipke. Clark Sr. wanted it because it was private, large enough for his family and, most importantly, sat on thirty-five acres. There, he built a racetrack and a barn where he could keep a stable of race horses.

O'Brien said that her grandfather provided an idyllic setting at her grandparents' house, where they spent days riding horses and tractors around the farm. Her aunt, Wyn, would drive the kids out there after church most Sundays. Although her grandmother Marie never liked the location—it was "too far out in the county," recalled Doug—her grandfather thrived there. O'Brien remembers how one barn was devoted to tool and die equipment so he could invent and experiment.

One horse, in particular, would make him famous. Clark Sr. got what newspaper reports called a "once-in-a-lifetime bargain" when he purchased a yearling for about $1,000 from the Walnut Hall Stud Farm in Lexington, Kentucky, in 1950. That trotter, Sharp Note, would go on to win multiple races in 1952, including the $24,000 crown in the Horseman Futurity, the $66,000 Kentucky Futurity and the $87,000 trophy in the Hambletonian, a prestigious harness race akin to the Kentucky Derby. The top prize in those three races earned Sharp Note the Triple Crown of harness racing. No Michigan horse had previously won any of these races.

News of the win earned headlines in magazines and newspapers across the United States, mostly because its rider was seventy-four-year-old Bion Shively. Shively, a veteran of the Spanish-American War who was known as the "Oklahoma Sage," was the oldest driver to win a Hambletonian, earning his victory on Sharp Note. The win even inspired a short film, *Old Man in a Hurry*. It was Clark's first showing at the Hambletonian; at this point, he only had been involved in harness racing for six years. Clark took home of a purse of $47,236.

Even in victory, Clark Sr. was good natured. Following a bet with family friend and physician Dr. H. Marvin Pollard, Clark Sr. donated half of his winnings to the University of Michigan Hospital. That money, some $20,314, went to equip Dr. Pollard's research laboratory. There, Dr. Pollard was conducting "investigations into gastro-intestinal and stomach diseases" in a search for an ulcer cure. Clark Sr., Dr. Pollard and President Harlan Hatcher attended the lab's opening, which included a tribute to Sharp Note. The horse's portrait was hung there "as a reminder of his contribution to medicine," the *Michigan Alumnus* wrote.

Clark Sr.'s semi-retirement to his beloved farm and race horses came after several incidents scarred his memories of the Dearborn Tool & Die, recalled Connie Norsworthy and her brother, Jamie Clark. The union was looking to form in the company, and some fellows caused trouble there on multiple occasions, Jamie said. Connie recalled a drive through the picket line: "I saw faces of men I knew that worked at the shop. They

Clyde W. Clark Sr. (right) holds the Hambeltonian trophy, which his trotter Sharp Note won in 1952 with rider Bion Shively (middle). Clark's wife, Marie, offers a congratulatory kiss. *Clark family collection.*

The Clark family spent much of their Sundays and summers at Clyde W. Clark Sr.'s retirement home near Ann Arbor and Ypsilanti. *Clark family collection.*

were screaming at us and screaming at the car as it went back to where the shop was."

"There were other episodes later on that led to roughing up some of the shop workers and other threats," Jamie recalled. "My grandfather finally said if he couldn't run his own company the way he wants, then he would retire from the tool-and-die business. He turned the business over to his sons and stayed in the shadows as the company's adviser and counselor."

Clark's three sons all held leadership roles at Dearborn Tool & Die. James, the eldest of the three Clark brothers, served as the vice-president. He was in charge of sales and public relations, was the mediator-arbitrator and assisted Clyde Clark Sr. in keeping the shop running, Jamie recalled. Connie remembers her dad being called the "peacemaker" of the family and business.

James was drafted in World War II, Jamie recalled, but the draft board considered him to be more beneficial to America where he was and deferred his enlistment. At that time, Tool & Die was helping the war effort by building military weapons.

Connie said that her dad was a fine dancer and actor. Jamie recalled his father as "a great tease. He loved life and had a fantastic sense of humor. He had a laugh that just filled the room, was a great dancer and had a beautiful tenor singing voice. He was in the Barber Shop Quartet in Dearborn and performed in many festivities in Dearborn, as well as for the Dearborn chapter. He was a Freemason, in which he was a thirty-second degree."

James met his wife, Lucille, at the Grande Ballroom, which was not far from the Dearborn-Detroit border, recalled Jamie. His father loved to dance, so he was a regular there. They would go on to have two children: Constance (Connie) and James (Jamie). "Dad was immediately attracted to my mother, who had a likeness to Lucille Ball with her naturally strawberry blond hair. My mother and her two sisters, Ruth and Gracie—the three Stomback sisters—were beautiful. I remember someone telling me my mother would not give my dad the time of day, but after two to three follow-up attempts by Dad, Mom decided he was a pretty good guy," Jamie said.

James's time at the Ford-Wyoming would be limited. He was diagnosed with Hodgkin's lymphoma, a type of cancer originating from white blood cells called lymphocytes. After a five-year battle, James died in 1953 at the age of thirty-nine. Had he gotten the disease today, modern medical treatment likely would have allowed James to live a much longer life. His death was said to draw the already close-knit family even tighter. "The funeral procession for my dad was very long. There was a police motorcade from the funeral home in Dearborn to Cadillac Memorial Cemetery in Westland, Michigan. We were later told this was one of the longest funerals in the history of this area. Many of the police officers that attended my dad's funeral were also employed at the Dearborn Tool & Die Company as part-time help to keep ends met for their families," Jamie said.

Clyde Clark Jr. served as the Tool & Die's treasurer. He graduated from Fordson High School as the salutatorian of his class, Connie recalled. He attended the University of Michigan to study law before joining the U.S. Navy, where he served as an officer. His ranking was high enough that the family never learned what exactly he did during the war. Clyde's daughter, Lynn, said that her father was a code breaker for part of the war and helped as a defense attorney. Clyde had had four children: Lynn, Clyde III, Pamela and Martin.

A great talker, Clyde perhaps had the the coolest temperament in the family. "Clyde had a very good business head and knew the numbers necessary to make quick and accurate decisions. When unsure, he always brought in advisers and sought other opinions," Jamie recalled.

He also was a caregiver, especially looking after his brothers, Lynn said. One of her favorite memories of her father and Harold was that they had lunch every day together, something that shows the strong bond between the brothers. "My dad was a really hard worker. He really believed in taking care of his employees, his family and his extended family. He was a good, caring person. I wouldn't trade in my family or my childhood for anything in the world," Lynn said.

Harold Clark served as senior vice-president. He also served in the U.S. Army during World War II, and his interest in the war and the military was a passion that would span his lifetime, according to his daughter, Diane. Harold attended Henry Ford Community College, where he studied business. "He was always interested in learning new things," Diane said. "He was an avid book reader and had an extensive library at his home in Dearborn. He was especially interested in World War II, German history, economics, American literature, vocabulary, photography, golf, Hollywood stars, U.S. presidents and health and wellness."

His granddaughters remember a man who was a true "night owl" and loved chatting. Carolyn Clark O'Brien Chang recalled their many trips to Hollywood and how much they loved meeting movie stars. "My grandparents were like Hollywood royalty to me; they always looked like they came off the set of a Hollywood studio," said Chang, who noted that she never once saw her well-groomed grandfather in blue jeans.

The Clark family frequently camped on family vacations. Here is Marie (middle) with her three sons, Clyde, Steven (right) and Harold (left). *Clark family collection.*

"I will always remember that my grandfather was a passionate seeker of knowledge independently and through learning from the experiences and knowledge of others," said Jennifer O'Brien. "He was an impeccable listener, and I believe this is one of the reasons he was successful throughout his life. There wasn't a time he wasn't there for any of his family for a simple hug, a lighthearted conversation or to give some advice."

After World War II, Dearborn Tool & Die was going strong, and the nation was hopping. The approaching decade, known as the Nifty '50s, would be marked by strong consumer spending, low unemployment, a "baby boom" and a movement from the cities to the suburbs. In 1955 alone, Americans purchased 7.9 million new cars, almost four times as many as a decade earlier.

By the end of the 1950s, four out of five Americans had a car, which symbolized "status, freedom and personal identity," according to *The American Dream*, a book about the decade. About one-fourth of all homes standing at the time had been built in the 1950s. And one-third of all Americans lived in what became known as "Suburbia."

Dearborn experienced a surge in housing starts during the late 1940s and well into the 1950s. The city was growing rapidly—its population was 63,589 in 1940 and would jump to 94,994 in 1950. The population would peak in 1960 at 122,007 people. Demand was high for cars, houses and entertainment for the many young families now filling the city.

Dearborn Tool & Die was the day-to-day responsibility of the Clark family. *From left*: Steven, Clyde Sr. and Clyde Jr. *Clark family collection.*

Throughout their lives, the Clark brothers would have heard about outdoor theaters known as the drive-ins. The first recognized drive-in opened on June, 6, 1933, in Camden, New Jersey. The R.M. Hollingshead Jr. Theater, named after owner Richard Milton Hollingshead, showed *Wife Beware*. Hollingshead knew that he had a hit and had already applied for a patent on his new ramp system.

Detroit got its first drive-in theater at 19440 Harper Avenue in Harper Woods in 1938, just four years after the concept was invented. Initially called Drive-in Theater, it later would be renamed the East Side. Tickets cost about thirty-five cents, and its first movie was *The Big Broadcast of 1938*, a Paramount Pictures film featuring W.C. Fields and Bob Hope. Despite a slight rain, attendance was said to be fair, according to *Box Office* magazine.

The East Side, also known as "Detroit's First Open Air Auto Theatre," would quickly become popular. Its advertisements promised patrons that they could "[d]ress as you please…Smoke…Laugh…Take refreshments…without disturbing others." Its screen, which measured about fifty by fifty feet, was considered the world's largest at the time. Its scale was such that "some of the ushers…have so large a section to patrol that they are using bicycles," reported a *Detroit News* article.

Not all residents were so enthused. In July 1938, John H. Flancher filed in court a petition with five hundred signatures alleging that this new drive-in theater was a public nuisance. Flancher claimed that its speakers projected sound so loudly that the movie could be heard from as far as two miles away. He asked for a restraining order against the drive-in. Judge Sherman D. Callendar adjourned the case so he could visit the theater himself. However, the court ended up dropping the case when the theater and Flancher settled amicably.

By some estimates, drive-in theaters were a $3 million industry by 1941. One year later, the concept of drive-in theaters was starting to expand across the United States. At the time, there were ninety-five drive-ins across twenty-seven states. Construction slowed dramatically through the war years, with only six drive-ins built. In fact, many theaters closed for months or even years during this time because of gas rationing and other shortages.

By 1946, there were about 155 drive-in theaters around. Within the next decade, the drive-in would experience its greatest growth period. Its biggest boom would occur after 1949, when the U.S. Supreme Court ruled that the drive-in principle could not be patented, preventing its New Jersey creator from forcing new exhibitors to license their basic ramp-parking system through him. By 1958, there were 4,063 drive-ins in operation.

Michigan had about 12 theaters around 1948, noted Gary Ritzenthaler, who tracks drive-in movie history on his website Water Winter Wonderland. That number would jump to 111 drive-in theaters in Michigan by 1956. (The state's number of drive-ins peaked at 137 in 1972. That total would drop rapidly in the decade to come as real estate prices jumped and owners sold their land for, in some rare cases, millions of dollars.)

Those early drive-ins were impressive, but the ones that followed would grow more and more elaborate. One of the most memorable was the Gratiot Drive-in, which opened on April 30, 1948, with a car capacity of 1,050. Not only did this Associated Theaters movie palace have kids' pony rides and a place to warm baby bottles in its restaurant, but it also had perhaps one of the most outstanding marquees in Michigan movie history. Its 115-foot tower formed a "living curtain waterfall, where 1,500 gallons of water a minute, illuminated by colored lights, comes rush down to duplicate Niagara Falls," the *Detroit News* reported.

Drive-in theaters were a symbol of America as a land of plenty, historians note. "It's like the ancient Roman Coliseum from an aerial and scale standpoint," said Robert Thompson, director of the Bleier Center for Television & Popular Culture at Syracuse University. "When you look at great folk heroes, the British have King Arthur. Who's ours in America? It's Paul Bunyan. What's his claim to fame? He's really huge!

The Clark family moved Dearborn Tool & Die from its original downtown Dearborn location to 10200 Ford Road, where it remained until the 1980s. *Clark family collection.*

That's perfect for Americans. This is the land of the Big Gulp. We like everything to be supersized."

This and other Detroit-area drive-ins were true reflections of the area's nickname as the "Automotive Capital of the World," Ritzenthaler said. "The economy then was good, and there was a high demand. People were spending money on drive-ins that would be unheard of now. For example, the Miracle Mile Drive-in in Pontiac cost more than $600,000 to build in 1960—that would be like $3 million today. It had projectors for 70mm film, and there were hardly any prints coming out in 70mm then. But that was the confidence people had in the drive-in industry."

A June 20, 1949 *Time* magazine article described the potential growth of drive-in theaters as phenomenal. "Within the year, U.S. drive-in theaters had doubled; more than 1,000 sprawled under the sky in 15 states, and with at least 100 more on the way, the sky seemed the limit."

Outdoor theaters were becoming more and more profitable as well. The *Wall Street Journal* reported that outdoor theaters grossed $126 million in 1949—a number that would grow to $271 million less than a decade later. Drive-ins accounted for 10.5 percent of all admissions in 1949, increasing to 26.0 percent by 1959. Concessions were a sizable profit center as well, going from $15 million in 1949 to about $108 million by 1959.

It made sense, then, that the Clark family would see the growth opportunities in drive-in theaters. They worked daily next door to a huge, empty field that had little potential for development. Sometime in 1948, the Clark family would pick up that chunk of land that used to hold a brickyard and then an airfield. They would begin their quest toward opening an entertainment venue they could call their own.

"My grandfather was a visionary and was a great believer in family unity," recalled Jamie. "The drive-in…was a way that families and individuals could escape from their workforce and enjoy recreational pleasure. My dad and uncles were all on the same page and supported my grandfather with his decisions."

The first mention of such a project showed up in the *Dearborn Independent* on April 29, 1949, when James A. Clark appealed to the Dearborn Zoning Board to "erect an outdoor theatre on Ford Road between Miller Road and Wyoming Avenue." The property at 10400 Ford Road, which was then zoned as "Industrial A," would need a Certificate of Compliance to become a drive-in. The zoning board approved Clark's appeal in May 1949.

However, the Clark family's victory wouldn't last long. They had petitions, parents' groups and a particularly pugnacious mayor named Orville Liscum Hubbard to battle next. Hubbard, who had been mayor of Dearborn since 1942, was about to start a very public fight against the Clarks' dream of opening a drive-in theater in "his" city. And if there was anything one could say about Mayor Hubbard, it was this: when he started a fight, he was always the last man standing.

CHAPTER 3

THE HOT POTATO

I'm all alone
At the drive-in movie
It's a feelin' that ain't too groovy
Watchin' werewolves without you.
—Olivia Newton-John, "Alone at a Drive-in Movie"

Arguably, the land next door to the Dearborn Tool & Die had been unproductive for years. Putting up a drive-in theater on what had been a largely empty field made perfect sense to the Clark family. However, there were more than a few Dearborn residents who disagreed. One in particular would make their lives miserable.

Despite approvals from the city's zoning board, the spread of drive-ins into the city of Dearborn drew the attention of two groups that opposed its construction. Up until this point, most drive-ins were positioned in rural areas; the Ford-Wyoming was breaking new ground by building within a city. The first round of protests came from a consortium of neighborhood clubs called the Dearborn Federation of Civic Associations as well the East Dearborn Kiwanis Club. Both of which went to the Dearborn City Council on June 14, 1949, to protest the proposed venue.

City Attorney Dale H. Fillmore told the city council that the Clark family needed a building permit to operate the theater. However, newspaper articles from the time show that the city council was unwilling to take a strong stand on the issue. According to the *Dearborn Guide*, council members "quickly dropped the matter" when the Clark family's attorney, John J. Fish, challenged the proposed building permit "and warned that the matter would be taken to court if the permit is withheld."

The Dearborn Federation of Civic Associations—which included at least eight neighborhood associations—was a name to be feared. The Federation was highly effective in determining zoning and development policies in Dearborn during this era. These activists "constantly lobbied local government on land-use issues," wrote David M. Freund in his book *Colored Property: State Policy and White Racial Politics in Suburban America.* The Federation asked the city council for everything from a stricter noise ordinance to revising zoning ordinances on what defines a "family" to rid the city of rooming houses. "[These groups] constantly made their preferences known to the City Council and city clerk. They wrote letters about sewers, bus stops, and weed ordinances. They requested that municipal garbage trucks be fully enclosed, complained about dirty, trash-filled lots and alleys, and identified traffic problem spots," Freund wrote.

By mid-June, the Dearborn City Council had changed its stand, announcing it did not intend to grant a license for the proposed Ford-Wyoming Drive-in. By now, ten individuals and community groups were protesting the project and filing objections with city council. Complaints came from the Dearborn Council of Church Women, the McDonald School Mothers Club and the North Dearborn Civic Association. Among the individuals listed were Dr. Edmund W. Waskin, a school trustee-elect of 5605 Michigan Avenue. Dentist Harold J. Lynch also was named in the June 21, 1949 *Dearborn Independent* article.

Fillmore said that the city had advised the Dearborn Department of Public Works not to issue a building permit for the outdoor theater. Council president Patrick Doyle noted to the audience that the Clark family had not yet applied for the license, so the council would not act at the present time. The "hot-potato case of the proposed Drive-in theater" had changed again by June 29 when the *Dearborn Guide* reported that the Clark family had filed a lawsuit against the city. A hearing was set for July 1 in Circuit Court to compel the city to issue a building permit, the newspaper noted, listing Clyde W. Clark Sr. as the site's owner.

By this time, the Clark family felt compelled to tell their side of the story. In a full-page ad in the *Dearborn Independent*, the Clark brothers decided to tell the "Truth About Drive-ins" and declared that "the citizens of Dearborn had the right to know [about] the new Drive-in now planned for construction on Ford Road near Wyoming." James, Clyde and Harold

Above: Mayor Orville L. Hubbard governed the city of Dearborn with a tight yet capable fist from 1942 to 1978. *Dearborn Historical Museum.*

Clark authored the introduction, noting that they were longtime city residents with "a manufacturing concern in Dearborn, founded in 1932." They wrote that they "have the best interests of Dearborn always in mind," and so they asked "an outside, impartial organization to make a thorough study of drive-in theaters and give us a written report." This report, prepared by "Leading Authorities," called the drive-in theater a "benefit to the community."

They ended their introduction with a strongly worded appeal: "We believe that you will overwhelmingly agree with us that this fine, new, modern Drive-in we plan will be a splendid addition to Dearborn's wholesome recreational facilities."

The report's authors from Detroit advertising agency Holden, Clifford, Flint Inc. described the drive-in theater as "accepted as an important and permanent contribution to the American way of life." They noted how these entertainment venues served as "one more extension of the usefulness of the family automobile," contributed to "the unity and stability of the family by making it possible for parents of very young children to attend the movies" and brought the "wholesome diversion of motion picture entertainment to thousands of elderly people, handicapped persons and others who have heretofore been classified as shut-ins."

In fact, the ad goes so far as to tie the potential success of Detroit's automotive industry to the future of drive-in theaters: "Any development that helps increase the use of the automobile helps the Motor City region and its citizens. The popularity of drive-in theaters, along with the growth of other facilities that encourage the use of the auto, such as shopping by auto, banking by auto, etc., is a good sign for the automotive industry...More autos used means more autos sold; more autos sold means more prosperity for the people of the world's auto capital."

Whether this advertisement gained the Clark family new sympathizers is unknown. However, Dearborn newspapers continued to report on the concerns that "flow into the offices of mayor and council," the *Dearborn Guide* wrote. These concerns focused, it stated, on "objections to the theater" based on "traffic, morality and undesirables."

The traffic jams would accompany "considerable expense to the city in added police traffic officers and would necessarily enhance the values of adjacent properties," George W. Donaldson told the council. Tobias Vanderweele, president of the Aviation Property Owners, added that "the location is too close to the city and for the betterment of Dearborn this should not be allowed."

For those who disliked the drive-in theater's proclivity toward becoming a passion pit, emotions flared up during the meeting. "Immorality in those places runs high," stated John Misiolek. "Unwholesome recreation for our youth," described McDonald School Mothers Club president Mrs. Vera Fox.

Other protests focused on the drive-in's appeal to "outsiders." Given that Dearborn's population for decades was nearly 100 percent white, this concern was not all that surprising. Dearborn had 35 black residents out of a 1940 population of 64,000, according to U.S. Department of Commerce data. It also came just a year after Mayor Hubbard successfully defeated a $25 million proposal to build a multi-family housing project in the city.

In 1948, the mayor had objected to a one-thousand-unit rental housing project by Chicago-based John Hancock Life Insurance Company on an estimated 930 acres owned by Ford Motor Company and the Ford Foundation. Although there was no proof that it was designed for or would attract African Americans, Mayor Hubbard had fought the housing development in part because of his longtime slogan "Keep Dearborn Clean," which many interpreted then and now as his desire to segregate the city from potential African American residents.

Right: Mayor Hubbard was a tough adversary who opposed the Ford-Wyoming Drive-in from its inception. *Dearborn Historical Museum.*

The "outsider" issue would come up time and again with the Ford-Wyoming. R.A. McPhea, president of the North Dearborn Civic Association, told the council: "The location will invite the element from outside Dearborn who will go to such places for reasons other than to see shows."

"We do not believe," wrote the Dearborn Council of Church Women, "this would prove a desirable addition to our city. We feel that it would tend to lower the morale of the community and might bring in undesirable persons."

Mrs. Emma Huth took a bolder stance: "If you let them get the permit, we will have all kinds of riff-raff in Dearborn. It is a disgrace to our city. Those places are not for people to see a clean show, but for our young people to sit in the cars and neck."

By July 20, 1949, Circuit judge John V. Brennan had decided that the drive-in theater did not need an operating license prior to Dearborn issuing it a building permit. The city via its Assistant Corporation Counsel Earl Smith said that Dearborn would appeal the ruling to the state Supreme Court.

The Clark brothers were the voice of the drive-in theater. *From left*: Steven Clark, Harold Clark (who served in the U.S. Army) and Clyde Clark Jr. (who served in the U.S. Navy). *Clark family collection.*

By now, those same community and civic groups were circulating petitions against the drive-in. At least one concern about this protest came from a letter to the editor in the *Dearborn Guide* that was published on July 27, 1949. As much as these groups were against the Ford-Wyoming, a few people were coming forward to voice their seeming lack of concern.

"A party came around to my house and asked me to sign a petition against somebody building a drive-in theater in the city. I told him I thought drive-ins were all right and I didn't care to sign the petition. When I said that, he looked at me as though my head was floating off into the air and couldn't believe his eyes," the author wrote. "I didn't much care one way or the other whether the theater was built until that guy with the petition came around. That made up my mind. Now I'm for 'em."

The issue pressed into summer and then the fall as the state Supreme Court gave the city more time to file its appeal. Finally, in November, the Dearborn City Council took an aggressive stand. All those present for a November 14 meeting voted unanimously to approve the building license for the Ford-Wyoming despite the case before the state Supreme Court. A *Dearborn Independent* article ended with an ominous note: the license approval had a two-week period during which time the mayor could veto its actions. And with Mayor Hubbard at the helm, rumors quickly began to circulate about his "five-point opposition" that would come at the next city council meeting.

By December, his concerns were well known, reported the *Dearborn Independent* in a December 2 article: "Hubbard declared the erection of an outdoor theatre on the proposed site would create a traffic problem, contribute to the delinquency of minors, attract undesirables, become an eyesore to the surrounding neighborhood, 2,000 residents of the area having petitioned against it."

The Ford-Wyoming Drive-in was located next door to the Dearborn Tool & Die shop, allowing the Clark brothers to work at both locations during the day. *Clark family collection.*

The Ford-Wyoming's main tower or screen is considered one of the largest in Metro Detroit. *Clark family collection.*

The Clarks reacted with a full-page advertisement in that same newspaper. Going point by point, the Clark brothers declared in their open letter to Mayor Hubbard that none of his so-called concerns held merit. This advertisement was even more emotive. The headline stated, "Please, Mr. Mayor! Don't Veto Wholesome Entertainment for the Plain People of Dearborn!" In this open letter to Mayor Hubbard, the Clark brothers addressed each of his worries, but they took particular umbrage in his description of the drive-in as an attraction for delinquents and undesirables:

Any charge you level against the Drive-In you must also level against the automobile. If you are against Drive-Ins you are against the automobile. You are against the means of livelihood for the majority of Dearborn's citizens. Furthermore, if Drive-Ins do contribute to delinquency, what will you do about it? Put up a wall to keep Dearborn youth from attending Drive-Ins outside of town? Or would you rather have a clean, well supervised theater right here in town? One that you and other parents can view with tranquil minds, knowing it is constantly under control and supervision of Dearborn's own police and health authorities?

They signed it with a final message: "It takes a big man to be honest enough and courageous enough to change his mind. We believe, Mr. Mayor, that you are that kind of man."

He wasn't. Mayor Hubbard indeed vetoed the license at the next city council meeting on December 6, 1949. The *Dearborn Guide* quoted him as saying, "Dearborn does not want to be a guinea pig city for the first drive-in theater within the city limits in the United States." He also said that the theater would present "a stock-yard appearance" with its high board fence and that it would become a "physical and moral liability to the community" that would not offset its prospective tax revenue. Mayor Hubbard even went so far as to compare the drive-in theater to the Hancock housing project, saying that "it is neither needed nor wanted by the people of Dearborn" any more than that controversial development, "and for much the same reasons," according to the *Dearborn Independent*.

Ultimately, it wouldn't matter, as six councilmen voted unanimously to override the mayor's veto, granting an operating license to the Ford-Wyoming. One councilman, Homer C. Beadle, even called the veto "ridiculous" and "a grandstand play," considering that this was the first time the council had a request for a license for a drive-in theater.

The council's decision may have come, in part, from the 9,500 signatures in favor of the drive-in theater filed that night. It also may have been political revenge—three of those six councilmen had been defeated in the November election and would leave office in January. Mayor Hubbard noted that the license "should not be rushed through a lame duck session less than four weeks before the new Council takes office."

Councilwoman-elect Lucille McCollough also chastised the council, speaking on behalf of community and mothers' groups. "Before the election, you would have no part of the drive-in, but immediately after the election you grant the license," she said in the *Dearborn Independent*. "I have been to some of

A side view of the main tower as it was being constructed on the former Haggerty Brickyard site. *Clark family collection.*

them and know what goes on in some of the cars parked in these drive-ins. I don't want that to happen in Dearborn."

The debate within the council chamber was described as "lively," touching off "mixtures of cheers, boos and applause." Those in favor of the drive-in theater who spoke at the meeting were paraphrased as noting that "people who go to outdoor theaters for immoral purposes could just as easily go elsewhere."

The next week, Mayor Hubbard "teed off" against city council for overturning his veto, according to the *Dearborn Guide*. The mayor went so far as to allege that one councilman received campaign contributions from the drive-in builders. "I think the council sold out," Hubbard said, "and not for a very high price—maybe a permanent pass to the drive-in theater."

The city's corporate counsel canceled its state Supreme Court appeal. The city issued the building permit in December 1949, and its construction began in earnest. By January, the land had been graded, and sewers were being installed, the newspapers reported. In April 1950, the theater was well on its way to its grand opening, a moment heralded by its architects, contractor and suppliers in another full-page advertisement in the *Dearborn Press*. Within it, nearly three dozen Ford-Wyoming vendors saluted the drive-in theater, noting their pride in being part of "making possible this modern means of entertainment that all Dearborn citizens and their neighbors may enjoy for many years to come."

Opening day was set for May 19, 1950. All was looking well: Brown's Bun Bakery would supply all of the hot dog rolls. Dearborn Tobacco & Candy Company would fill the shelves with treats and cigarettes. The Detroit Popcorn Company supplied the popcorn machine and supplies. Advertisements in the newspapers promised that it was "the newest and finest drive-in theater in the Detroit area."

The first movie was to be *The Man from Colorado*, starring Glenn Ford and William Holden. The plot promised "Lawless Violence!" Admission was set at sixty-five cents per adult. Children under the age of twelve were admitted free.

But before the screen could light up and the neon be turned on, the Clark family would have another fight on their hands. The city struck back—and at the last minute. On May 18, new Dearborn fire chief Frank J. Gilligan accused them of failing to install a fire alarm, fire hydrants or a proper passage for fire department apparatus in case of an emergency. Chief Gilligan also termed the ten fuel storage tanks and lack of hydrants as a "distinct fire hazard," according to the *Dearborn Independent*. He recommended the city investigate and withhold the theater's certificate of occupancy.

Those fuel tanks didn't belong to the Ford-Wyoming. Rather, they were the property of the neighboring Socony-Vacuum Oil Company. However, a state law prohibited the presence of fuel storage tanks within 300 feet of places where the public is likely to congregate. One of the company's tanks was said to be only 130 feet away from the last row of cars that would be parked within the theater.

The city had a "Stop Work" order issued, and it was placed on the theater. The Clarks' attorney threatened court action, and it was taken down. A

The Clark family owned the concession stand, which was unusual for drive-in theaters at the time. It proved one of its greatest financial assets. *Clark family collection.*

The projection booth sits back from the main tower, allowing for the proper distance from the screen. *Clark family collection.*

certificate of occupancy was issued reluctantly. Jamie Clark recalled that Mayor Hubbard was present that opening night and had a ribbon placed across the entrance of the drive-in to prevent its opening. "My grandfather took down the ribbon and gave Mayor Hubbard a booklet of season passes to the drive in. The opening went on as scheduled, and to my knowledge Mayor Orville Hubbard never attended future events at the drive in," Jamie said.

The Clark family had finally had enough. Although the state fire marshal was set to come in for a look of his own, Clyde Clark Jr. was quoted in the *Dearborn Independent* saying that the show, as it were, would go on as scheduled. "We have complied with all the laws of the City and State. Now let them try and stop us. We beat them in court on their own terms; our theater is ready for opening and we invite the citizens to inspect it and pass judgment on it," Clyde told the newspaper. "This is just one more attempt to deprive decent citizens of their legal rights in the community."

On May 24, 1950, a group of state officials, including representatives of the State Fire Marshal's Office and the Attorney General's Department, gave its approval to the newly erected drive-in theater at Ford Road and Wyoming Street. Clyde Clark Jr. was vocal once again. "This whole thing is nothing but politics. The state fire officials inspected the theater six weeks ago and approved it. The city had nothing to do with it, but they just stuck their noses into the matter," the *Dearborn Press* reported him saying.

The Clark family wasn't done there. On May 25, a front-page correction ran in the *Dearborn Guide*. Within it, the editors apologized for not contacting the theater for its article about the fire department's concerns. And it also apologized for using the phrase "fire trap," which it said was "unwarranted under the circumstances."

The battle was won. The Ford-Wyoming Drive-in theater, for all intents and purposes, was open for business.

CHAPTER 4

A NIGHT AT THE DRIVE-IN

So I just stared though the door screen and I
watched the cars come down the pike
their lights against the sky
like a drive-in movie on a country road
that I've seen before.
        —*Catherine Britt, "Drive-in Movie"*

The Ford-Wyoming's opening night on May 19, 1950, was marred slightly by a spring storm. After all, it is hard to see a drive-in movie in the rain. But the crowds didn't seem to mind; they easily filled the parking lot. It was a sign of things to come—cars full of families, teenagers and young lovers would line the theater's driveway in the years that followed.

The drive-in celebrated its opening with jubilation. The long, hard fight against Mayor Hubbard and city officials had been intense. The Clark family's joy at finally seeing those first cars roll in was palpable. In one of its earliest advertisements, the Ford-Wyoming welcomes patrons with seemingly endless enthusiasm: "Now Open! Closest to You! At your doorstep! Newest! Most Modern! Most Centrally Located! The latest RCA equipment throughout!"

The first movies were a mix of family-friendly pictures, popular comedies and rough-and-tumble cowboy classics. When *Road to Rio* left, *Desperadoes* replaced it. *My Friend Irma* featured a *Tom & Jerry* cartoon for the kids. Abbott and Costello got them laughing with *Pardon My Sarong*, with a Popeye short.

GRAND OPENING TONIGHT

THE NEWEST AND FINEST DRIVE-IN
THEATRE IN THE DETROIT AREA

LAWLESS VIOLENCE!

FORD   HOLDEN

The Man from Colorado

DRFW

A COLUMBIA PICTURE

FORD-WYOMING
DRIVE-IN

Ford Road and Wyoming Avenue. Reached via Ford
Road, Wyoming Avenue, or Willow Run Expressway.

Admission: Adults, 65c; Children Under 12 Free

The Ford-Wyoming opened for business on May 19, 1950, with *The Man from Colorado*, drawing crowds to the young drive-in. *Clark family collection.*

Fans packed in for Rita Hayworth in *Cover Girl* and Victor Mature in *Red, Hot and Blue*.

The Clark family went out of their way in these early years and throughout their tenure at the Ford-Wyoming to open the facility up for nonprofit groups and charitable organizations. They would host shows for groups ranging from the Dearborn Torch Fund to the Kiwanis Club, helping them raise money or celebrate events. One such fundraiser for the Kiwanis got attention from *Billboard* magazine, which reported on the event as a "precedent for outdoor theaters." The Kiwanis had its members serve as the ushers and ticket takers, while the theater provided cashiers for the well-attended fundraiser for National Kids' Week.

Getting the theater up and running—along with taking care of their growing families—occupied James, Clyde and Harold Clark. Harold, the youngest, would end up taking on the drive-in as his personal interest, although all three still held positions and responsibilities at Dearborn Tool & Die.

Having Harold at the helm also would be a natural fit because of his theatrical nature, his family recalled. He enjoyed dancing, and he met his wife, Nathalie, at a community dance at Detroit's Grande Ballroom, whose second-story dance floor made the live music venue on Grand River one of the hottest spots in town. Harold also was the family singer, even entering amateur contests and making a few recordings to show off his fine voice. His daughter, Diane Clark O'Brien, recalled him singing continuously when she was a child.

O'Brien, who was born in 1953, said that the daily life of running the drive-in was truly a family affair. Brothers Doug and Steven were born in 1958 and 1962, respectively. All three kids worked at the Ford-Wyoming, doing everything from selling popcorn to cleaning the bathrooms to working the driveway so the cars made it into the rows in an orderly fashion. Those jobs stepped up during the summer months. For example, Doug remembers being paid one dollar per hour to refinish the hundreds of speaker poles with Rustoleum paint. "That's where I learned to drive a car. It's where I met my first girlfriend," Steven Clark recalled. "I grew up there."

O'Brien said that her father would drop her off at school in the morning and then head to the Ford-Wyoming, where he would go over the previous night's paperwork in the theater's wood-paneled office. Harold would tend to the financial matters of the facility, working with the secretary (likely Grace or Dorothy) in her nearby office. This is where Clark Enterprises, the company that owned the theater, operated day to day.

The marquee was an eye-catcher. Note in the background that the tower does not yet have its extra sides at the top for Cinemascope. *Clark family collection.*

The Ford-Wyoming with its iconic lettering. Much of the decorative elements that added its Art Deco charm have yet to be added. *Clark family collection.*

For the most part, Harold was the Ford-Wyoming's nervous system. He understood the pulse of the audience. What did they want to see? How could he match two different films to make them a popular crowd pleaser? Steven recalled how much his father loved recording the list of movies that the theater would show, giving it a flair and sophistication because of how much he enjoyed the task.

The crowds enjoyed life at the drive-in as well. James A. "Jamie" Clark recalled how the Ford-Wyoming's proximity to Henry Ford's Rouge plant made for great shows during the show. "As a special treat, there was often the experience of observing the Ford blast furnace off in the distance as we watched the movie on the big screen," Jamie said. "The sky would light up, and we could hear the *oohs* and *aahs* around us as we watched the movie."

Another time, Harold was able to get the army to bring in a Sherman World War II tank and park it next to the concession stand, Jamie recalled. "I remember Harold saying it was difficult putting the gravel back in the parking lot in the areas where the tank would make its turns. It was a publicity stunt that was successful, and people loved it," Jamie said. "I also remember standing alongside the tank and admiring its size and power."

Attendance would soar during celebrity sightings. Movie and television stars often did promotional tours at that time, attending screenings of their films and doing meet-and-greets with fans. Diane recalled seeing Herman's Hermits at the Ford-Wyoming, doing publicity for their MGM Movies, such as *When the Boys Meet the Girls* (1965) and *Hold On!* (1966). Another of her favorites was Dwayne Hickman, who portrayed the teenage heartthrob on the CBS television show *The Many Loves of Dobie Gillis.*

Not every star proved to be so polished. Doug Clark remembers the night rock-and-roll singer Jerry Lee Lewis came through. Harold, who tended to be a bit conservative, didn't enjoy that visit as much as the others, his son recalled. "Jerry Lewis came in a motor home and stayed for maybe ten minutes. He seemed like he was half in the bag," Doug laughed. "He stood on the projection booth, and my dad never liked him after that. He felt [Lewis] had an attitude."

Another time, Harold managed to get his hands on one of the hottest tickets in Detroit: the Beatles at Olympia Stadium. He advertised that he would be giving away a set of Beatles tickets in the drive-in's popcorn, bringing in fans by the droves. Clark held back four seats for his family to go see the show in September 1964, but they ended up staying for only two songs because his mother wanted to get away from the chaos and screaming girls, Doug recalled.

Working with film distribution companies took up the majority of Harold's day at the Ford-Wyoming. Negotiating the percentage of profit that the drive-in would earn through the distribution companies was his dad's favorite (and, in some instances, least favorite) task, Steven remembered.

"Generally, the film company would get 80 percent, and my dad would get 20 percent. It would depend on if it were a first-run film or not. If the film were slightly older, the split would be more like 70/30. To bug my dad to get a really cool movie—one that had just come out—was nearly impossible. He would complain about how he didn't want to pay the 90 percent that some films had, like anything with James Bond in it. But he would do it…sometimes," Steven said.

Because the drive-in theaters had to share revenue with the film companies, distributors would keep a close watch on how many cars went through the gates, Doug and Steve recalled. "The film company would have a 'counter' or 'clicker' sitting across the street, keeping track of how many cars came into your lot. After the film, you would have to call in your car count to an 800 number. It would have to match what the clicker got within a reasonable number. And you never knew when they were watching," Steven said. "You would have to keep those ticket stubs for something like five years. We'd have lunch bags, and we'd put the tickets in them. They were bagged every night and dated."

Watching how much the drive-in spent on everyday items was a major part of running the business, Steven recalled. For example, the in-car heaters were a luxury item that the Ford-Wyoming promoted in its newspaper advertisements. However, Harold and theater manager Boyd Beauchene only allowed them to be turned on occasionally because they were expensive to operate. Plus, a few patrons would always find a way to steal them, although they operated on a voltage level that didn't work with typical residential systems.

Keeping the neon lit was another battle, Steven recalled. "When the neon signs would go out, I would tell my dad about it, but he'd always say it was too expensive to replace. So I'd go to my mother next," Steven chuckled. "She'd start to work on him. 'Harold, you've got to fix the neon.' And he'd do it if she asked. When I got old enough, I'd just call Downriver Sign myself and got it repaired."

As evening fell, Beauchene would arrive to take over and get the theater ready for the nightly show. Beauchene was Nathalie Clark's brother, and he ran the Ford-Wyoming like it was a military operation, Doug and Steven remembered. Every detail was considered, and nothing was left to chance.

Harold Clark would serve as the theater's primary caregiver, working most mornings in its wood-paneled offices. *Clark family collection.*

The concession stand carried everything from candy to popcorn and beef sandwiches, one of its best sellers. *Clark family collection.*

He would arrive at about 5:00 p.m. to turn off the security system and turn on the bun warmers, as well as other concession-stand equipment. "If it wasn't for Boyd, the theater wouldn't have been what it was," said Doug, who worked alongside his uncle until the family sold the theater. "He was instrumental in every aspect of the drive-in."

Diane agrees. "Uncle Boyd did everything. If it wasn't for him, my dad wouldn't have been successful. Boyd gave up his whole life for the drive-in." And Beauchene had the respect of nearly everyone who knew him, Steven added. Although he was a quiet man, Steven recalled how people of all walks of life would come in to chat with Beauchene as he stood behind the counter. "It was a well-run operation. He was always there on time, and you could always count on him," Steven said.

The ushers arrived next. They were typically teenage boys who lived near the Ford-Wyoming. They started to show up at about 6:00 p.m. Their first task was to turn off any speakers that had been left on from the night before, Steven recalled, because Beauchene didn't want them blaring as customers began to arrive. The ushers typically dressed in everyday street clothes and wore reflective vests to help guide traffic, assist patrons and serve as playground attendants. They also were the ones who drove around the property and sprayed mosquito repellent to keep the grounds relatively bug-free. "The ushers became our friends. They were our age. They all came from that area: McGraw, Michigan Avenue, southwest Detroit," Steven said.

In its early years, the Ford-Wyoming had a large playground in front of the screen. Families could come into the theater's grounds as early as 6:30 p.m. so their kids had time to play on the toy train, spin on the merry-go-round or swing. One of the most popular attractions was the red motorized boats. Because they sat in an above-ground pool, they needed constant supervision. That was one reason why this attraction was among the first to go, ending up as play equipment in the Clarks' backyard for most of their childhoods, they recalled.

On the side, the ushers and the Clark brothers would keep an eye out for any money lost the night before. Steven said that they would determine which way the wind had been blowing and then walk the theater's fenced perimeter to search for any dollars that might have blown out of a patron's hand or on the way to their pockets. Typically, they would find anything from singles to twenty- and fifty-dollar bills along that route, Steven said.

The next task was to take inventory. This was, by far, one of the most difficult and time-consuming jobs for the person working that night at the concession stand, Harold's sons recalled. Beauchene was so meticulous about keeping track of concession stand items that any leftover popcorn that

The field had enough parking for more than four hundred cars. Each car had its own speaker pole. *Clark family collection.*

The neon that lit the front gave the drive-in theater the glamour of a Hollywood movie palace. *Clark family collection.*

went home to the Clark kids had to be sent in an old Brown Bun box rather than waste a usable one, Diane noted.

"We never had a cash register. It was all done out of a till with coins. The bills went into the box. So the only way we knew what we made at night was beginning inventory—which was taken the night before—and the ending inventory. Every night, you had to count every candy bar, every cup, and every popcorn box was accounted for. You didn't leave that night until your drawer was balanced," Steven recalled.

The concession area was one of the Ford-Wyoming's most important profit centers because the Clark family owned the stand, O'Brien said. During nights when popular films were showing, Harold would add movie shorts or cartoons to give patrons more time to get to the concessions area and back to their cars, Doug recalled.

Throughout the night, the staff stayed in contact with one another to post updates, Steven added. "Every thirty minutes, you called on the intercom to the concession stand to the ticket booth to get the car count. You did that all the way to closing time to determine the needs of the concession stand," he said.

For all of Beauchene's prep work, a few calamities happened, like when the films would show up late. "There were a few times we didn't receive some of the movies until thirty minutes before the show would start. Uncle Boyd would panic if they hadn't shown up," Steven recalled. The movies arrived in an unmarked truck. "Because they'd come in overnight air, they'd still had the airline tags on them and everything. They'd arrive in these big, octagonal steel boxes that were big and heavy. The delivery guy would have a key to our Edison room. He would put the two pieces in the Edison room, then we would take them to the projection booth. We'd prep them, put them on the reels and splice them together." Then there were the moments when a projectionist made a mistake. "Spectators would just blast the horns whenever something happened in the projection booth that interrupted the movie," Jamie recalled.

Every Thursday, Steven would change the marquee sign, that colorful, neon-lit masterpiece that sat at the entrance of the theater. Its huge size served as a beacon for the Ford-Wyoming. "It was one of my favorite things to do. We'd do it late at night. We would put a ladder up on the catwalk and climb up there. You'd put the actors' names and which picture they were in at the time. Cars would drive by, honking and waving at you," Steven said.

It was a good thing that Clark and his sons were taller men as well—not only did it make the marquee job easier, but their heights were also useful for intimidating those carloads of teens that tried to sneak into the theater.

Harold Clark was a sharp dresser who adored Hollywood movies and its stars, particularly Errol Flynn and Clark Gable. *Clark family collection.*

You always could tell which vehicles had hidden passengers in the back, Doug and Steven said. "You'd have a car come in with handprints all over the trunk. The guy would have pizza and beer in the car. And you knew he wasn't coming by himself. We'd radio it back to the ushers. The guy would be foolish enough to park in the back row, get out of the car and open the trunk. Then, we'd charge them [for admission]," Steven said.

"If they were nice, we'd let them stay. If they were mean, we'd charge them and then make them leave. We kind of had a mean streak in us,"

Steven laughed. "Sometimes, they'd come over the fences by twisting the barbed wire. Others would try to sneak in by coming in the exits at a high rate of speed. We always caught them."

Doug added, "I've actually seen where they've taken the spare tire out of the trunk so they can fit in there. And the tire would be sitting in the back seat."

Even family members would try to sneak people in. "I was fortunate enough to be a blood relative, so I was advised I could have as many in the drive-in as I could get in one car for free," Jamie recalled. "But I was told, 'This stuff has got to stop' when I was trying to get three or more cars full of my friends in. I guess I got out of hand."

Mostly, people paid their admission. But there were a few occasions where the Clark family would take these scofflaws to court. "One time, we saw this car come in, and its trunk was riding low so we knew there were people in it. We followed the car around, and we confronted them when they finally let the people out of the trunk. They started to give us a hard time, but the police were sitting right there and asked us if we wanted to issue them a ticket. So we said yes," Doug said.

"We had to go before the local judge, and those guys told the judge how it was a hot summer night. The judge started to yell at us, saying how those guys could have died in there. We told the judge that if they were dumb enough to get into the trunk, that was their problem," Doug said. "The judge thought about it a bit and said, 'Well, guys, this is going to be the most expensive movie you never got to see.' And they had to pay for the ticket."

After the movie was over, huge floodlights would illuminate the field. The ushers, along with Doug and Steven, would go around and wake up any sleeping patrons. They would clean up a bit and then head home. Or not. "I remember when CB radios were big in that day. Doug didn't like ladders, so I got some coax cable and ran a CB antenna all the way up on the tower," Steven said. (Doug noted, "It was terrifying, going to the top of the screen. That's why I don't like heights at all.") "When we'd close at 3:00 a.m., we'd plug the cord into the CB radio, sit and talk to girls or people who were still up around the country. Boyd never knew we did that; he would have thought we were nuts," Steven laughed.

It was, in hindsight, an idyllic albeit unusual childhood. "If you had a car fire, a few people come over the fence or a car sneaking in through the exit and it was a marquee-change night, I was happy. It didn't get better than that," Steven said.

Families may have dominated drive-ins when they first opened, but teenagers ruled the roost during the 1960s, noted filmmaker and film

Left: Boyd Beauchene was the Ford-Wyoming's night manager, running the drive-in with military-like precision. *Clark family collection.*

Below: The Clark family also owned the Town-N-Country Bowling Lanes in Westland. Harold Clark is standing on the left with a vest and spats. *Clark family collection.*

historian Gary D. Rhodes, who conducts research in the areas of early cinema, documentary film and the horror film. At the Queen's University Belfast, Rhodes is the co-director of film studies and lectures on such topics as "Hollywood Cinema" and "Cinema and Memory." His books and anthologies include *Horror at the Drive-in* and *White Zombie: Anatomy of a Horror Film.*

Because many drive-ins had to battle to get first-run films from distribution companies, lower-budget filmmakers found room to show their works there, Rhodes said. Horror film directors, in particular, embraced this venue, creating movies for teens and with teens as the primary actors. That drew more young people to the drive-in who turned it into their personal hangout space, Rhodes added.

"The baby-boom generation took over the drive-ins like they did the movie-going audience in general. It became the chance to get away," Rhodes said. "You get this incredible place for teenagers to hang out. They've bought a ticket, so they have a legal right to be there and it's safe enough. You could talk to people without someone shushing you. You could walk around, congregating…There's no other experience like the drive-in experience."

Although the Ford-Wyoming was their baby, the Clark family would go on to open another entertainment venue—the Town-N-Country Bowling Lanes in Wayne, Michigan. Jamie Clark recalled how much of an entrepreneur his grandfather was for the times. "I remember being in his basement in his Ypsilanti farm while he was sitting at his drafting table, which I have to this date, and observing him drawing on paper extending onto to the floor and asking him what he is doing. He modestly replied, 'I'm going to build a bowling alley, Jimmy,'" Jamie recalled. "It was nothing but country out there at that time, but my grandfather said, 'Everything is moving in this direction, and people are going to want recreation.' Like his vision with the drive-in, he saw a need for a bowling alley out in the middle of nowhere."

The Clark family would own the bowling alley the longest of any of their enterprises, selling it in 2004. But the Ford-Wyoming always stood out as the favorite. "I think out of all of his businesses, my dad really liked the drive-in the most. He liked the movies. He liked the movie stars, and he got to meet many of them. He liked the socialization. He just really enjoyed it all," Steven said.

CHAPTER 5

UP IN FLAMES

We both fell sound asleep
Wake up little Susie and weep
The movie's over, it's four o'clock
And we're in trouble deep.
—*Everly Brothers, "Wake Up Little Susie"*

A lthough life in many respects was sweet, running the Ford-Wyoming was a huge enterprise. There were challenges, incidents and accidents—such as the 1967 Detroit riot and a 1979 tower fire—that would change the way Harold Clark and his family looked at the drive-in business.

The biggest challenge continued to be Mayor Orville L. Hubbard. Even if the Clark family may have moved on from the building license debacle, Mayor Hubbard clearly had not. The drive-in was still on his radar, a place few people wanted to be. Although things at the Ford-Wyoming were running smoothly, Hubbard would have at least one more trick up his sleeve.

Right around the time of the Ford-Wyoming's first anniversary, Mayor Hubbard decided that the drive-in wasn't paying the city enough in terms of its operating license fee. He brought up the fee in April 1951 with the council, saying that the license fee did not balance the costs of having Dearborn police officers at the theater to help with traffic.

Hubbard contended that the license fee of fifteen dollars monthly was "poor business" when the drive-in was charging sixty-five cents per person for admittance, according to an article in the *Dearborn Guide*. Councilman Joseph

Ford countered this argument, noting that the "owners of the drive-in were substantial taxpayers, residents and were therein entitled to public service."

The mayor lost when the council voted to support the Ford-Wyoming and the Clark family. The *Dearborn Guide* led the story of his defeat in this way: "Mayor Hubbard's annual attempt to block the license of the Ford-Wyoming Drive-in met with failure Tuesday evening." However, the fee was increased to $500 annually to help cover police costs.

The song was the same in 1952. This time, Councilwoman Lucille H. McCullough stepped up to do battle with the drive-in. Although the city had signed off on the operating license, the feisty Dearborn councilwoman wanted a fight. She asked the city council to force the theater to change its hours, noting her opposition to the 3:30 a.m. closing time. According to McCullough, "The late hour contributed to the delinquency of Dearborn's youth." She suggested that 2:00 a.m. would be more reasonable. The council tabled the issue, and the operating license once again was issued.

Why would the mayor and his minions show such derision for the Ford-Wyoming? What had the Clark family done to earn his displeasure? Chances are, they did nothing wrong—that was just the way Hubbard operated, said David L. Good, a lifelong Dearborn resident and author of the definitive book on the mayor, *Orvie: The Dictator of Dearborn*. "He was very charming when he wanted to be and when he was with people he wanted something from," Good said. "But he also could be an absolute son of a bitch to those who worked for him or were on his wrong side. He was unrelenting."

The Ford-Wyoming likely was a convenient scapegoat for Hubbard's frustration with his own personal troubles, Good said. "This was such a turbulent period for him," he added. "He was being besieged on so many fronts during this period that he may have just figured this was something he could take the high moral ground on and relieve some of the pressure from the attacks he was under."

To understand Hubbard, you have to go back to when he first took office. Dearborn was in the midst of public scandals and police corruption when Hubbard ran for mayor of the city for the first time in 1941. He had been trying to gain public office for nearly ten years at that point, and this was his first taste of power. (He'd run for state senator four times, the city council once, Dearborn justice of the peace once and U.S. Congress once.)

Generally, Hubbard used his charisma and chutzpa for the betterment of the city—he kept the streets clean and the parks groomed. Dearborn city services operated like clockwork. He built Camp Dearborn in Milford as a retreat for city residents. He'd work from 7:00 a.m. to midnight with no complaint and welcomed anyone to call him or stop by his office. He sent out birthday cards and other greetings, signing them with his customary green ink and in inch-high lettering.

But he also dominated just about all facets of city government for nearly four decades. According to an August 21, 1950 article about Hubbard in the *Detroit Free Press*, the portly politician was a force of nature: "With the help of a 'strong man' city charter pushed through by his supporters, Orville chopped the once-dominant city council down to size, hired and fired city department heads at will...An ex-autoworker and ex-Marine who started with nothing and worked himself through night school to a law degree, Orville Hubbard brooked no opposition."

Reader's Digest, which also did a lengthy profile on Hubbard in September 1955, gave a laundry list of the mayor's many nicknames, including Tyrant and Brainy Opportunist. "I've known him for 20 years. He's done more good than bad for Dearborn and he's kept the town in a turmoil. That's the way he wants it," the magazine quoted George St. Charles, then editor of the *Dearborn Independent*.

Bill Clark, one of the Ford-Wyoming's current owners, said that he briefly knew Hubbard toward the end of the mayor's last term. He described Hubbard as the "Pied Piper": "He was charismatic; he could get people to follow him anywhere on earth. Whenever he was around, everyone lined up to follow."

At the time the Ford-Wyoming was being built, Orville Hubbard's life was, for all intents and purposes, a tangled mess. His wife, Fay, was suing him for divorce, complaining that he gave her only ten dollars per week to run the house, abused her and belittled their four children. Such allegations made front-page news in all of Dearborn's newspapers.

Also, city reformers had started a recall campaign to boot Hubbard out of office. Some council members had accused him of bullying and passed an ordinance prohibiting him from coming in contact with city residents, for which he would face a $500 fine or ninety days in jail.

And John J. Fish, the Clarks' attorney and a longtime Hubbard detractor, had hit the mayor with a $100,000 libel lawsuit for his 1950 election hijinks. The trial centered on a 1949 handbill that Hubbard distributed that claimed that "John Poisson" was part of a Dearborn police scandal and tried to extort money from the city. (As an aside, Hubbard also was known for saying, "If anybody beats up John Fish, cripples him up, does anything to get him out of my way, such a person can have any job he wants.")

The trial judge in the Fish libel case called it the "most viscous language in Michigan case law" that he'd run across in libel proceedings, Good said.

A furnace fire in 1979 caused massive internal damage to the drive-in's main tower, burning holes through its screen. *Clark family collection.*

Fish won the libel suit, and Hubbard refused to pay the $7,500 fine that the court determined. So Hubbard would end up going on the run, setting up a temporary office in Windsor, Canada, to avoid Fish and the fine.

His combativeness at this time extended beyond political antagonists, Good added. Hubbard ordered Fire Chief Stanley Herdzik to chop down Henry Ford II's office door and make trouble, find violations and get bad headlines for Ford Motor Company. Ford himself, Good said, was pumping money—as much as $100,000—into the Hubbard recall effort.

So, amid his own chaotic life, calling the Ford-Wyoming a place where morality would be in jeopardy was classic Hubbard, Good said. "That kind of blatantly phony, self-serving description of the drive-in echoed one of his statements in the crossbill he filed in his divorce. It was something like, 'There is something far greater and of a more lasting quality than mere temporary pleasures, especially those which are of a sensual nature.' You can see him dictating that and chuckling as he did it," Good said. "He'll say anything and say it memorably…If you were a civilian or if he didn't

The front of the drive-in shows how the fire moved upward from the furnace through the main tower. *Clark family collection.*

have a political ax to grind, he was very, very charming. But if you were an antagonist, you were in big trouble."

With such a fiery opponent, other problems probably seemed small to the Clark family. Interestingly, none of Harold Clark's children recall him saying anything about Hubbard, good or bad. As was their custom, the Clark family rose above personal attacks, noted Connie Norsworthy. "You'd really have to bait them to get them going. They weren't interested in controversy," she said. "But they didn't let anybody take advantage of them, either. They were very supportive of one another."

Even some of the steel beams that support the main tower were damaged in the 1979 fire, family members said. *Clark family collection.*

However, the Ford-Wyoming wasn't immune to bad times. The Clark family battled occasional crime, such as people sneaking into the theater. But they also had several break-ins, Harold's family recalled. Thankfully, they ended up being minor incidents over the three decades that they owned the Ford-Wyoming. "One time, someone cut a hole through the ceiling in the office. They got in, but they didn't get anything," Steven Clark recalled. "Another time, our alarm company called, and I went down there. It was in the winter sometime, and there was freshly fallen snow on the ground. The guy walked to the concession stand and projection booth, where he tried to get in. Stuff was disturbed, but nothing was really taken. When the police got there, they followed his footprints in the snow. They walked right up to his house."

Because the Ford-Wyoming sits on the border of Detroit, its neighbors' issues affected the drive-in in myriad ways. One of Detroit's most troubling events came on July 23, 1967, when a violent public clash between Detroit residents and city police resulted in a deadly altercation that some called a civil disturbance. Others called it a riot.

The 1967 riot is said to have begun when the Detroit police raided an unlicensed, after-hours bar or "blind pig" at the corner of Twelfth and Clairmount Streets. The back-and-fourth between police, patrons and people on the street resulted in a brawl that unleashed a five-day riot—one of the most destructive in U.S. history. President Lyndon B. Johnson sent in U.S. Army troops, and Michigan governor George W. Romney brought in the Michigan National Guard to help quell the violence. In all, 43 people died, nearly 1,200 people were injured and some 7,200 were arrested.

Controversial Detroit mayor Coleman A. Young described that week's events in his book *Hard Stuff: The Autobiography of Mayor Coleman Young*:

> *Detroit's losses went a hell of a lot deeper than the immediate toll of lives and buildings. The riot put Detroit on the fast track to economic desolation, mugging the city and making off with incalculable value in jobs, earnings taxes, corporate taxes, retail dollars, sales taxes, mortgages, interest, property taxes, development dollars, investment dollars, tourism dollars, and plain damn money. The money was carried out in the pockets of the businesses and the white people who fled as fast as they could. The white exodus from Detroit had been prodigiously steady prior to the riot, totaling twenty-two thousand in 1966, but afterwards it was frantic. In 1967, with less than half the year remaining after the summer explosion—the outward population migration reached sixty-seven thousand. In 1968 the figure hit eighty-thousand, followed by forty-six thousand in 1969.*

Former Dearborn resident Sharon Carden Brown said that she was at the Ford-Wyoming on the night of the 1967 riots. It was a memory that lingers to this day for Carden Brown, who was seventeen when the incident happened. "They turned off the movie after the first hour and said to go directly home," she said. "I remember hearing that night that Mayor Hubbard had snipers on rooftops, and we saw the burning on the news when we got home. I also remember being mad that we weren't give a voucher for another time. They were just in a hurry to get us out of there."

Doug Clark also recalls those nights and the chaos. "People actually took a couple shots at the oil tanks" next to the Ford-Wyoming. "Thank God nothing happened."

Perhaps the most dramatic accident at the drive-in theater came in January 1979, when a fire started during a night when the Ford-Wyoming was closed. Steven Clark recalled the fire came from a runaway furnace, which started running and couldn't turn off. It ended up sparking a fire that was so hot it warped the theater's steel beams.

Harold's son-in-law, Tim O'Brien, was working at Ford Motor Company and going to night classes at the Detroit College of Law (currently Michigan

The front of the tower received minimal damage from the fire. *Clark family collection.*

Opposite: Damage to the drive-in's front was extensive and required significant repair work. *Clark family collection.*

State University Law School). As he drove by on his way to class, he noticed the fire. "Working full time, going to law school, married to Diane and having our first daughter, Carolyn, approximately three years old at the time, was a very busy time in my life. I would leave Ford at 5:00 p.m. and drive directly to DCL. There was no time to even have dinner," O'Brien recalled.

"My usual route took me east on Ford Road, right past the Ford-Wyoming theater, to the Wyoming entrance to eastbound I-94. I remember casually looking out my window in the direction of the theater, and I could see black smoke coming from the top-right corner of the screen tower," O'Brien said.

O'Brien knew how quickly that screen—which was nearly three decades old and had layers upon layers of paint on it—could go up in flames. "At that time of day, it was unlikely that any of the employees would have been there. Fortunately, the Dearborn Tool & Machine Company offices were right next door, so I quickly turned into that parking lot, ran into the office and saw Clyde Clark," O'Brien noted. "Clyde came right over and started to say how nice it was to see me. I interrupted him and said the theater is on fire—call 911! I had to leave for school."

Repairs to the front of the tower required new panels to be installed. *Clark family collection.*

The new panels on the tower front are still visible on this iconic building. *Thomas Tomchak.*

O'Brien continued, "I never got to have the nice conversation that Clyde wanted to have or see what became of the fire. These were before the days of cellphones, so I couldn't even call Diane and tell her what had happened until I got to a pay telephone at school. I really believe that the timing and coincidence of me driving by the theater at the early stage of the fire, and then being able to quickly get to Clyde, probably saved the theater from a catastrophe."

Despite O'Brien's early intervention, the damage was severe. Holes could be seen on the front and the back of the Ford-Wyoming's iconic screen. Harold took a memorable photo from inside the screen itself, looking out onto the children's merry-go-around covered in snow. Work began quickly to repair the theater, which would reopen that spring.

If you look closely at the front, you can see where the building was fixed. The panels are a slightly different color than the original, creating a kind of zigzag pattern in the façade. Its once pristine face now began to show its age. And the constant efforts of keeping the drive-in in shape and to the standards that the Clark family wanted may have been starting to take its toll.

Plus, times were changing. The roving bands of teens that once kept the Ford-Wyoming hopping were starting to hang out at malls and other venues. People found larger indoor theaters more compelling, perhaps drawn by the improvements in their screens and sound. Some movie directors even vocalized their dislike for drive-in screens, noting that their films didn't look as good when projected on an outdoor theater.

Steven recalls how the rise of videocassette recorders and subscription television were moving into the movie industry's territory: "I was twenty when I got the chicken pox, and I spent hours watching movies at home through our On-Subscription television box. It was this little box that sits on top of your television and a dish that sat on your roof. That was the first version of cable television."

Harold's longtime friend and occasional dinner companion Charles Shafer had been working on him for some time now, asking more and more frequently whether he was ready to sell the Ford-Wyoming. Now, it seemed, Harold was finally ready to listen to an offer.

CHAPTER 6

THE DEANS OF DETROIT

Got my first kiss in the back seat of a car
At the drive-in, the drive-in movie show
And I gave her my heart
I was in love right from the start.
                    —*Steve Miller Band, "Circle of Love"*

Charles Luther Shafer grew up inside a movie theater, cleaning up after crowds, running the concession stand and selling tickets alongside his mother, father and siblings. The Shafer family's investments in indoor and drive-in theaters resulted in a movie palace dynasty that dominated Wayne County.

The Ford-Wyoming was the final and most long-lasting jewel in their crown. Its purchase would come to symbolize the end of the family's era in many ways, as changes in property values and interest levels would result in the sale of most of the Shafers' properties.

Much like the Clark family, the Shafer story starts with a clever patriarch. Walter Dennis Shafer's empire included nearly a dozen theaters and spanned four decades. Walter was considered one of the nation's pioneers in the theater construction and management business. But it was an industry that had its highs and lows. One reporter described Shafer's movie houses as an enterprise that "lost him one fortune and made him two others."

Walter Shafer was born on October 20, 1885, in Indianapolis, Indiana. He was working as a public stenographer when World War I began. He began promoting war bonds, selling them to audiences across the country. His natural flair for working the crowd helped him succeed in these "pep talks." More importantly, it introduced Shafer to the stage and movie theaters.

His theater career started when he began managing multiple theaters, including the Fox Liberty Theatre in Elizabeth, New Jersey. He moved to St. Louis to manage its Fox Theatre there and then Detroit in 1921 to work at the Fox-Washington Theatre, which was located on the northwest corner of Washington Boulevard and Clifford Street. The grand theater was part of the theater chain created by William Fox, the New York native best known as the founder of movie studio giant 20th Century Fox.

Walter Shafer moved over to its sibling, the Fox Theatre, a mega movie palace that opened in 1928 and still stands on Detroit's Woodward Avenue. He left there to become the general manager for the Woodward Theatre Corporation, which was the state's largest independently owned chain with twenty-seven theaters.

Headed by Henry S. Koppin, the Woodward Theatre Corporation managed theaters that were popularly known in the Midwest as the Koppin Vaudeville Circuit. Many big acts that toured Detroit, such as Al Jolson and Fred Astaire, also played the Wayne. However, the financial impact from the stock market crash of 1929 caused Koppin to close his chain of theaters.

The Great Depression hit Metro Detroit like a hammer. Theaters began to close rapidly due to dwindling crowds and concessions. Walter had an eye for opportunity, so he negotiated to buy the shuttered Wayne Theater from the Woodward Theatre Corporation. Walter Shafer borrowed money from his mother and purchased his first movie house for $5,000, the equivalent of $100,000 today, Charlie Shafer said.

"My dad loved the theater business," Charlie recalled. "He bought the Wayne for dirt cheap because nobody was buying anything. But he knew the business, so he could run it. My mother was the cashier. My father ran the projector. I took the tickets. My brother took over the candy counter. My brother and I would clean the theater afterward. So we didn't have any payroll."

Within two years, Shafer and his young family were struggling. Some poor investments, along with the theater's meager attendance, had them on the ropes. They had to close the theater several nights a week because no one was coming. Some nights, they'd make as little as seventeen dollars, as Walter Shafer told a newspaper interviewer.

Likely in desperation for audiences, Walter Shafer created a marketing ploy that would ultimately save him and his business: he offered the "10 Cent Night," when every adult got into the movies for half price. Charles

Shafer told the *Wayne Dispatch* that those Thursday nights were epic, filling the theater to the point that the police had to stand on watch because most of the town was inside to watch the show. Audiences slowly grew to the point that the Wayne Theater was getting as many as two thousand people a night. By 1939, Shafer had built up the business up and saved enough pennies to build the Shafer Theatre in Garden City.

At their peak, the Shafer family owned ten theaters: Wayne Theater (1930), the State-Wayne in Wayne (1946), Wayne Drive-in Theater (1949, closed in 1990), Algiers Drive-in theater in Westland (1956, closed in 1984), the Quo Vadis complex in Westland, Shafer–Garden City (remodeled and renamed the La Parisien) in Garden City, the Dearborn Theater in Dearborn (1941, closed in 2006), the Dearborn Drive-in in Dearborn Heights (1948, closed in 1984), Willow Drive-in in Belleville (1966, closed in 1983) and the Ecorse Drive-in in Taylor (1951, closed in 1989).

Charles Shafer told the *Detroit Free Press* in a June 21, 1996 article that drive-ins were entirely new to him in the 1940s and that the unique concept grabbed his attention. "I had seen a drive-in over on the east side. It was just

Opposite: Charles and Lillian Shafer (left), movie star Edward Everett Horton and Dorty and Martin Shafer (right) at the 1965 groundbreaking of the Quo Vadis Theater in Westland, Michigan. *Wayne Historical Museum.*

Left: Charles and Martin Shafer went into business together to operate the movie empire that their father, Walter Dennis Shafer, started in 1930. *Wayne Historical Museum.*

a field—half the cars couldn't see over the other cars. Then, I saw one in New Jersey and it had ramps."

Despite their excitement over opening their first ozoner, the Wayne Drive-In proved more difficult to construct that the family anticipated. Once built, the tower at the Wayne toppled. With the help of his brother, father, mother and a structural engineer, Shafer got the screen back up and a love affair with outdoor theaters had begun.

The family business changed when Walter Shafer died on November 1, 1961. To pay tribute, all of his theaters closed for an evening to honor the great movie man and his contributions to the industry, the local newspapers reported.

He left the business to his sons, Charles and Martin. Although they were known for their internal spats, the two were in agreement that there was room to grow within Metro Detroit's theater industry. "I had just gotten back from Army. My dad said you can sell the theaters or you can work them. He told us, 'You have to make a decision. But you just can't sit there. You either have to get bigger or get smaller,'" Shafer recalled.

The Shafer Wayne Theater in Wayne, Michigan, was the Shafer family's first movie house. *Wayne Historical Museum.*

The Quo Vadis was the Shafer family's most dramatic design, from Metro Detroit architect Minoru Yamasaki, who is best known for the World Trade Center. *Wayne Historical Museum.*

Charles and Martin stepped up to take over their father's empire. As the owners of Wayne Amusement Company, Charlie and Martin, with their mother's support, would manage his theaters as well as build up a portfolio of swanky movie houses, night clubs and entertainment venues.

As a team, the Shafer brothers earned nicknames like the "Deans of Detroit's theater men"—joining hallowed names such as the Sloan brothers and the Goldberg twins, as well as Lou and Nick George, all of whom owned and operated Metro Detroit's best and showiest movie houses.

Many consider Lillian Shafer, Walter's wife, to be the "Queen of the Shafers," establishing the over-the-top architecture and interior décor that made the family's theaters so memorable. Born in New York City on November 26, 1899, Lillian (Theimer) Shafer graduated from high school in Elizabeth, New Jersey, and studied business for a time before working as a bank teller. She met Walter Shafer in 1918 and married him a year later.

The Shafer family was best known for two theaters: the Quo Vadis and La Parisien. The Quo Vadis was a lavish affair—it was described as an "entertainment center" that included the Algiers Drive-in as well as four indoor theaters. One of its claims to fame was that the structure was designed by Minoru Yamasaki, best known for the World Trade Center in New York City. The Quo Vadis also included the Over 21 Club, a restaurant and cocktail lounge, one of the first to offer alcoholic drinks to its patrons. The club allowed people to see the Algiers' neighboring screen through a glass wall and listen to the show with headphones while enjoying the lounge's more sophisticated setting.

Martin Shafer told the *Associated Press* for a 1978 article that serving alcohol did not affect concession sales. "People like popcorn to go with their beer." And cleanups had been minimal because "scotch and soda doesn't make the mess that cherry pop does."

The Quo Vadis was "at the time…the best theater in the country," Shafer said, with seating for 1,200 people. The exterior was described as sleek and modern, yet the interior had a lavish Roman motif. Life-size statues of the Four Seasons decorated the lobby. Italian tile covered the exterior walls.

Charles and Martin also renovated the Shafer-Garden City to become La Parisien Theatre at Ford and Middlebelt Roads in 1964. Its name was inspired by its French provincial décor. It was lauded for its Bavarian crystal chandeliers in the auditorium, gold wallpaper with black flocking, marbled lobby with a recessed fireplace lounge, the first rocking theater chairs in Michigan and a heated marquee to keep theatergoers warm as they entered the building.

In a November 9, 1964 article, *Box Office* magazine described it as a "highly distinctive house" that "surpasses anything ever done before, and its

Charles Shafer and Harold Clark were personal friends, often dining together with their wives at the Dearborn Country Club. *Clark family collection.*

uniting of majesty and sheer architectural splendor forms the highest degree of moviegoing enjoyment."

Charles and Martin's partnership began to cool in the mid-1980s, as Martin was looking to retire and both of their parents were now deceased. As a result, the Shafer family began to divest its holdings. Shafer said they sold the majority of their indoor theaters to National Amusements Inc. in Dedham, Massachusetts.

Soon enough, many of the family's biggest treasures became more storefronts. The Algiers became Westland Crossing, a shopping center development with a Toys "R" Us. The Wayne went to Ford Motor Company to build the Wayne Assembly Plant. The Dearborn Drive-in was turned into a Farmer Jack. The La Parisien became a U.S. Armed Forces recruitment center. The Willow Drive-in became the Park Estates Mobile Home Park. The Ecorse Drive-in became a McDonald's restaurant.

Although these sales would prove hugely profitable, Charles knew that he never wanted to retire. Of everything he once had, only one theater would remain: the Ford-Wyoming.

CHAPTER 7

THE GANG'S ALL HERE

Katie and Tommy at the drive-in movie
Parked in the very last row
They're too busy holdin' on to one another
To even care about the show.
—Trisha Yearwood, "She's in Love with the Boy"

One theater remained elusive to the Shafer empire: the Ford-Wyoming Drive-in. Charles Shafer and Harold Clark were longtime friends, having met during the meetings that area theaters owner would have from time to time. Over the decades, the two would go to dinner together to talk shop or socialize with their wives. After all, when you've worked in the movie business your whole life, there are only a few people who have the stamina to stay up as late as these two did. "You have to enjoy having dinner at eight or nine o'clock at night. Because there's no such thing as an early dinner with Charles," said Bill Clark, Shafer's current business partner.

Shafer started asking Clark about the Ford-Wyoming during these dinners, waiting for the word on when he was ready to sell. Shafer said he also tried to buy the Grand River, but he couldn't convince that owner to budge. Shafer said that he liked everything about the Ford-Wyoming. Its location was particularly attractive. Having a theater in the city meant that you got first-run movies before the rural drive-ins, and that ensured a full parking lot. There was a Chrysler plant next door, and those autoworkers liked drive-ins because they stayed open long after indoor

Top: Charles Shafer purchased the Ford-Wyoming Drive-in theater in 1981 after years of asking Harold Clark about its sale. *Clark family collection.*

Left: Nathalie Clark was the Clark family's rock, providing unconditional love to her husband, children and grandchildren. *Clark family collection.*

theaters turned off the lights—plus they could bring in a few beers to drink during the show.

The Ford-Wyoming also sat at the border of Detroit, whose population was well over 1 million residents. Although there were about twenty-five other drive-ins around at this time, the Ford-Wyoming was the closet one to the city. That made for a captive audience, Shafer said. "If they wanted to go to a drive-in, they had to go to the Ford-Wyoming or they didn't go," he noted.

His years-old wait ended in 1981 when Nathalie finally took the initiative during one of their dinners with Charles at the Dearborn Country Club, where the Clark family was members. The Clarks' children say that their father was still on the fence, but his wife knew that the time had come. The Clark family also would sell Dearborn Tool & Machine Corporation in 1982. According to newspapers, Bloomfield Hills-based Newcor purchased it for a reported $7.1 million.

It makes sense to Diane that her father would have discussed the sale with his beloved wife, his partner in marriage and in the Ford-Wyoming. "My father, Harold, told me that his biggest asset in his life is his wife, Nathalie. Without her great personality and sense of business, he wouldn't have been as successful," she said. "She was a great asset during business dinners and was always a great influence on all that were lucky enough to know and love her. He always stressed that my brothers and I were very fortunate to have such a loving, giving, caring and beautiful mother."

The transition was handled with just a few simple sentences. "I think Harold's ready to sell the drive-in," Nathalie said, according to Doug and Diane. "Why don't you buy it off of him?" And with that one question, the Clark family's three-decade love affair with the Ford-Wyoming was over.

The two old friends—bound by their passion for drive-in theaters and the Ford-Wyoming in particular—shook hands, never signing any formal paperwork about the ideal. Shafer said that he offered Clark $1 million, and his friend took the offer with no negotiation. "Being in theaters, I was probably closer to him than I was to anybody else in the business. We were good friends for many, many years," Shafer said. "He loved that theater. The only reason he sold is I think he wanted to retire. I think I surprised him when I offered a price to him. I think he figured I'd tried to knock it down a little. I told my brother, 'I think we just bought a drive in.' He told me I was nuts."

One of Shafer's first missions was to find a businessman interested in splitting the operation. His brother, Martin, was lukewarm to say the least. So Charlie turned to Bill Clark, who had been working for the Shafer family at its theaters as a manager. Clark (no relation to the Ford-Wyoming's original owners) met the Shafer family when he was the proprietor of a bar and restaurant that was down the street from the Dearborn Drive-in. Martin Shafer, a regular there, asked Clark if he would be interested in working at the Quo Vadis at the Over 21 Club because of Clark's food-and-beverage background. Clark was soon running the entire complex as manager.

Working in theaters was ideal for Clark because he liked having two or more businesses going at one time. He could devote his daytime to one operation and his evening hours to the theaters. Clark agreed to become business partners with Shafer on the Ford-Wyoming, Willow and Ecorse Drive-ins, but the Ford-Wyoming was always the one that they all enjoyed the most.

"He and I have enjoyed a more than thirty-year relationship at the drive-in. We've never had a harsh word between us. He takes care of the movie bookings. And I take care of overseeing operations, maintenance and other various things. We've split up the duties pretty well," Clark said. "It's often a joke between us that we cover the clock twenty-four hours a day. My day starts at three or four [o'clock] in the morning and ends at five or six [o'clock] at night. That is when Charles is just getting out of bed."

Shafer was a natural at the theater's scheduling, Clark said, picking out the films that the Ford-Wyoming would show and for how long. Having the right double feature could fill the lot to capacity for several nights in a row. "He is the master of putting combos together of movies that draw the public. When he puts two together, it creates a blockbuster. It has that right synergy," Clark said. "There's no formula. There's no computer to work out the calculation. It's his natural instinct borne out of more than sixty years of experience."

Owning both ozoners and indoor theaters gave Shafer a particular advantage when it came to selecting which film to show. "The drive-in has always been a youth market—families and kids," Shafer told the *Detroit News* in 1983. "Because of that, I'm careful not to get too sophisticated in drive-ins. That audience wants action: shooting, blood, chases, noise…We used to murder 'em with the great cowboy and war flicks, but they don't make 'em like 'Dodge City,' 'The Fighting Sea Bee's or 'Guadalcanal Diary' anymore."

Errol Flynn movies, westerns or any war pictures also scored huge hits, Shafer told the *Free Press*'s Bob Talbert in 1994. "We'd run 'em for a year sometimes and pack 'em in." The top drive-in draws then were "Eddie Murphy in anything," Arnold Schwarzenegger and Sylvester Stallone "in the action films. Horror pictures and comedies, too. The Hudson brothers

The Ford-Wyoming began to add screens under the ownership of Charles Shafer and Bill Clark, a longtime Shafer employee and business owner. *Waterwinterwonderland.com.*

The drive-in added four more screens on Wyoming Road to capture more of the crowds that filled its parking lots every night. *Waterwinterwonderland.com.*

and Robert Townsend films. Whoopi Goldberg is a big draw. Black stars do extreme well in drive-in. Some movies we can play eight to ten weeks. 'Menace II Society' was No. 1 last year. We ran it almost four months."

Holidays and hot summer nights provided a boom in business as well. "Holidays are great. Richard Pryor's 'Stir Crazy' once sold out several showings on Christmas night," Shafer told Talbert. Another time, the Ford-Wyoming had to turn hundreds of cars away when it screened Prince in *Purple Rain* in 1984.

When asked about Bill Clark's high praise, Shafer wasn't shy in telling about his knack for picking out the right combination of films for the Ford-Wyoming. "I was so good that other theaters would call me from all over— Iowa, Nebraska," Shafer said. "Some of it is easy, like putting two John Wayne movies together. Something like *Flying Tigers* could pack them in. If you combined Laurel and Hardy with an Abbot and Costello picture, you'd get a mob...One time, I got lucky with putting together *The Days of Wine and Roses* with *Whatever Happened to Baby Jane*. That appealed to women."

Woe to the poor projectionist who went against Charlie's directions. "One time, we had a great combination of *Bambi* and *No Business Like Show Business*, and that dumb projectionist put *No Business* on first. I could hear people's horns going off, so I came running," Shafer recalled. "He said to me, 'What should I do?' I said, 'Stop the film and put the right one on first.' He never made that mistake again."

Working the crowd was another part of the job. If things got busy, both Clark and Shafer leave the office and help direct the cars that would spill out onto Ford Road and Wyoming Street. That was one of Shafer's favorite moments—"Mr. Showman," as Talbert called him, would smile and wave to all of the people as they drove past.

And they always came, no matter the weather. In fact, Shafer brags about how the Ford-Wyoming only had to close three times in the more than three decades he's owned it. "In the winter, when there's six inches of snow on the ground, we have 200 cars at the Ford-Wyoming Drive-in," Shafer told the *Detroit News* in 1985, reporting that the theater was sold out 75 percent of the time on Saturday nights.

Certain nights are always busy, noted Virgil Berean, the theater's longtime manager. Berean has worked for Shafer since 1976 and used to manage the Ford-Wyoming 6–9 Theatre. When those screens were demolished in 2010, he relocated to the 1–5 Theatre. Virgil started working as a projectionist in 1986 because "there was no one else who wanted to do it." "They come for Valentin's Day like there's no tomorrow. Doesn't matter if it's Tuesday,

Thursday or Sunday. They'll even show up even when it's freezing cold," Berean said.

Having a Chrysler automotive plant next door also was a boon for the Ford-Wyoming; that built-in audience of autoworkers is one of the reasons the drive-in could buck the overall trend and stay open seven nights a week, said Gary Ritzenthaler, webmaster of Water Winter Wonderland, an online resource about Michigan drive-ins and other entertainment venues. "Everybody tried it. But the Ford-Wyoming was the only place that kept on doing it. It was those guys who got off on the midnight shift. They'd hit the store and get a bottle. And then they'd catch a couple movies at the drive-in," Ritzenthaler said.

With regular crowds, Shafer and Clark took a hard look at the sprawling lawns that surrounded the Ford-Wyoming. At one time, the Clark family kept them so well manicured that the golfing family would play a round or two in front of the drive-in's screen. But with his construction experience, Clark realized that he had an asset that wasn't being appreciated. Bill Clark supervised the construction of the additional screens over the years.

Thanks to these new screens, the Ford-Wyoming could bring in additional movies to appeal to families looking for a cartoon early in the night and a blockbuster movie for mom and dad. Films that were doing well could stick around longer, boosting overall traffic and concession sales. Also, if one film sold out, patrons that showed up for that show likely would go to see another one instead of heading home. There was always something to watch at the Ford-Wyoming, and the long lines became a regular thing. "It's the key to survival in the drive-in business," Shafer told the *Detroit News* in 1983. "You've got to diversify, to get as much out of each location as possible. The more product you've got to show, the better your chances."

Starting in the mid-1980s and continuing through the next ten years, Ford-Wyoming grew and grew. Shafer and Clark snagged screens from now defunct drive-ins and brought them over. Screen 4 was built on the former front lawn. To add Screens 3 and 5, they purchased two contiguous parcels from neighboring businesses. The final screens came when Shafer and Clark leased a piece of property on Wyoming, adding four more to the complex.

As an aside, evidence of the brickyard is still clear on the Ford-Wyoming property. When Theater 5 was installed, the builders found what seems like a wooden basement there as they dug the posts into the ground, according to Manager Virgil Berean. That likely was an outbuilding of some sort for the Haggerty brickyard, he suggested. And bricks turn up across the property whenever maintenance or repair work is being done there.

Longtime Ford-Wyoming manager Ed Szurek was one of the theater's biggest boosters in the Metro Detroit community. *Kenny Karpov.*

Manager Virgil Berean came over to the Ford-Wyoming from another Shafer property and now serves as its caregiver. *Kenny Karpov.*

"The front lawn was aesthetically pleasing, but we agreed it wasn't making any money. It was what you might call a non-revenue producing parcel, so we decided to put a theater there," Clark said. With regular crowds, a good economy and open land, it made sense to expand, Shafer and Clark agreed. "Many of the indoor-theater companies in Michigan like AMC were taking their big theaters and splitting them into multiplexes in the late 1970s and early 1980s. What used to be a one-thousand-seat theater would be divided to make four two-hundred-seat theaters," Clark said. "So the idea to split the drive-in wasn't anything new; it was being done elsewhere in the country. We thought we'd better our chances for success if did that at the Ford."

The Ford-Wyoming's size is what made it a standout among drive-in theaters as a whole across the nation. At one point, it was said to be the largest in the world. "The scale of the Ford-Wyoming's operations was unusual. To go from a single screen to nine at one point is pretty rare within the drive-in community," Ritzenthaler said. At one point, the Ford-Wyoming could handle three thousand cars nightly, about double its original footprint. "It's the Godzilla of Detroit-area drive-ins," the *Detroit Free Press* noted in a June 14, 1991 article.

"We could expand the Ford-Wyoming because it's in a heavy industrialized area. That's also a problem. We can't expand again because there's no more land available around us. We went against the trend by adding a theater in '84, another in '85, '88 and '89 and three in '90," Shafer told the *Free Press*'s Bob Talbert in his June 4, 1995 column.

And the audiences always came. People of all walks of life found their way to the Ford-Wyoming. Young couples coming from the casino or beaches on Belle Isle would stop by. Taxi drivers who had just signed off from work would cruise in. The young guys who stocked the shelves at nearby grocery stores would pull in after midnight. Everyone was welcome at the drive-in, Shafer said.

Celebrities also made their appearances, although most chose to show up incognito, Shafer said. "One time, [fighter] Tommy Hearns showed up and said to me, 'Charlie, don't let anyone know I'm here.' I told him it would be a lot easier to do that if he didn't show up with three Cadillacs with his name all over them." Another regular was Detroit mayor Coleman Young. "He'd come in an ordinary car, and nobody would pay attention to him. He could see a show and not be bothered," Shafer said.

Life was good, and Charlie was right where he wanted to be. "If you like what you do, you'll never work," he told the *Wayne Dispatch* for a 2010 article.

CHAPTER 8

DETROIT DOWNTURN

Vacation under summer stars
Yeah, we'll hop in a jalopy to a drive-in movie
And never look at the show, we're gonna hug and kiss
Just like this and I can't wait to go, go, go.
—*Connie Francis, "Vacation"*

To outsiders, owning a drive-in theater may seem like a dream job. There is a palpable magic to the property despite its barren nature. Perhaps it is the tower's mass looming over the sky. Maybe it is the nighttime quiet that sets the mood. But in reality, there is little romance to the ownership and operation of an ozoner, particularly when the economy tanks, automotive plants close, your neighboring city begins its decades-long crawl toward bankruptcy and someone tries to sell the land where you make a good portion of your income.

Shafer owned five other drive-ins when he purchased the Ford-Wyoming, so he had experience with the cyclical nature of the business. After all, you cannot run an outdoor theater in Michigan—where temperature and meteorological swings are the norm—without understanding that things weren't always going to go your way. Plus, Shafer understood that outdoor theaters had already seen their prime. Having lived through the opening and closure of a number of them already, he knew that running the Ford-Wyoming would always have its risks. "The height of drive-ins was probably in the 1950s or 1960s," Shafer said. "After that, business slowed down. There are so many other things people could do to amuse themselves."

Shafer's era with the Ford-Wyoming saw two significant losses. The first was the closure of the neighboring Chrysler plant, which had drawn regular customers to the drive-in theater. The second was Detroit's sustained population losses and growing reputation for crime and other safety-related issues.

According to Bill Watson, curator of the Allpar website, the 6000 Wyoming Avenue plant was built in 1917 during World War I for arms manufacturing. In 1919, it became the property of Saxon Motor Company, a firm started by Hugh Chalmers. With the end-of-war economic downturn, Chalmers closed down Saxon. General Motors Corporation then purchased the plant, and it was used by Buick and Cadillac. Chrysler bought it in 1934. In 1936, the plant expanded and set up for assembly of DeSoto cars. DeSoto engines and bodies were trucked over from East Jefferson. Chrysler built the McGraw stamping plant just to the east; it made oil pans, valve covers and other small stampings.

In 1992, the Wyoming plant was demolished; a few years later, the McGraw plant was expanded. But in 1998, Daimler acquired Chrysler, and within two years, the writing was on the wall. In 2003, McGraw Glass closed. All of those loyal drive-in enthusiasts were gone, a loss that diminished the number of patrons at the Ford-Wyoming in noticeable ways.

Detroit's population loss started in the 1960s, but it picked up momentum after the 1967 riots. The city that borders Dearborn and the Ford-Wyoming saw its population drop from a high of 1.85 million in the 1950s to 1.20 million in 1980. Its decline would slow between 2010 and 2012, when a reported 701,475 people were living within Detroit's 139 acres. That slowdown didn't last long. In 2014, the U.S. Census Bureau put the city's population at 688,701, down nearly 10,000 residents from 2012. The city's suburban counties—Oakland, Macomb and Livingston—gained about 1 percent in population.

"What killed the city of Detroit was the riots in 1967. After those, Detroit completely changed…That was the end of [new movie theaters] in Detroit. After that, they only opened up in the suburbs," Shafer said. "People moved out. From that day on, the city's population went down and never stopped."

Losing half a million residents to the suburbs not only hurt the city, but it would also eventually kill the Ford-Wyoming's claim that it was the largest drive-in theater in the nation. A Florida drive-in with thirteen screens popped up to take that title. And with fewer and fewer people making the drive from Detroit or neighboring cities, there just wasn't enough revenue coming in to justify the Ford-Wyoming's screens in Theaters 6 through 9, Shafer said.

It didn't help that the property's owner also hoped to sell this parcel, putting up a "For Sale" sign that panicked longtime Ford-Wyoming fans that

Keeping the Ford-Wyoming in good shape takes constant maintenance. Replacing the lettering is said to cost more than $1,000 per letter. *Thomas Tomchak.*

the whole thing was shutting down. Shafer and Clark tore down the four neighboring screens along Wyoming in 2010, dropping the drive-in back to five screens that were on the property he and Clark owned outright. Without those screens along Wyoming, the theater began to call itself the Ford Drive-in on its website and in advertisements.

"Things were going along well until the economy went down in 2008. As the economy sank, so did the moviegoing business," Clark said. "Those years

between 2008 and 2012 were pretty bleak," Clark continued. "Our clientele is largely working families. They're mostly blue-collar workers and with automotive ties. During those years, there were massive layoffs combined with the collapse of the banking industry and General Motors' bankruptcy [in 2009]. All of those things were reflected on our business."

The Ford-Wyoming's longtime free admission for children was changed to two dollars for each kid. Rather than raise its adult ticket prices, creating this

The light poles that Harold Clark's sons used to paint each summer show some rust and age. *Thomas Tomchak.*

new admission fee put the drive-in in sync with indoor theaters, which were charging as much as six dollars per child at the time, Clark said.

There were other reasons people stopped going to drive-ins. These theaters were surrounded by challengers. The advent of television was hard enough, but most households started buying video gaming systems as well. Although cable television had been around since the 1950s, its true boom time began in the 1980s. The VCR, which had been introduced in the 1970s, became a staple in U.S. households. People could not only record popular sitcoms like *Rosanne* and *The Cosby Show*, but they could also rent a movie from the local video store and watch it at home.

The 1990s brought the Internet and a new attachment to personal computers. Cable companies began offering DVRs, which recorded television shows much like a VCR. The addition of smartphones and tablets further pushed people away from traditional theaters and drive-ins to a large degree. Why sit outside to watch a show when you could stream the latest release right from your iPhone or Android device?

Another drive-in killer was the real-estate boom. Most drive-in theaters were built shortly after World War II and into the 1960s. Property was relatively

One of the regular maintenance challenges at the Ford-Wyoming is replacing speakers, which people steal for personal use. *Clark family collection.*

To replace lost speakers, the Ford-Wyoming purchases ones from other drive-ins that have closed in Metro Detroit. This one is from the Gratiot Drive-in Theater. *Author's collection.*

inexpensive, especially if you built outside city centers or in rural areas. As property prices grew in the 1980s, drive-in owners who were nearing retirement started receiving sizable if not outlandish offers for the significant acreage they owned. Ozoners that once sat far outside town were now inside the city limits or were highly desirable for large housing or industry development.

"It takes fifteen or twenty acres to open a drive-in. You can't build a drive-in in a town. You have to go into what I call the boondocks. You get a big piece of property. But the city expands, and you become part of the city. That's when it becomes popular to sell," Shafer said. "Listen, if you bought the land for $50 an acre and someone offers you $5,000 an acre, you're not going to say, 'No,' for a business that might pay you $100,000 a year...When I sold my land to Ford, I sold it for ten times when I paid for it."

Added Bill Clark, "Most drive-ins sit on seventeen to twenty-one acres. That happens to be almost an ideal size for some of these shopping centers. So now if you're a drive-in owner, you're making a couple hundred grand a year and someone offers you $4 million for your land, you're going to take it."

It seems so strange that people's move to the suburbs now begins to threaten the drive-in theater, said Robert Thompson, director of the Bleier Center for Television & Popular Culture at Syracuse University. "The drive-in begins to thrive in postwar civilization. But the thing that makes the drive-in explode also is the very thing that kills it. Suburbanization largely kills the drive-in," Thompson said.

The Shafers also sold many of their indoor theaters in the 1980s because adding enough screens to make these older theaters competitive with new multiplexes just became too burdensome, Shafer explained. "Indoor theaters were one screen normally, especially if they were built from 1900 to 1950. That's how you did it—it was one screen at a time. All of a sudden, these newcomers got the idea to put up two screens, then three. Then it was fifteen! That's when I quit—it got too expensive," Shafer said.

For the most part, Shafer and his management team say that running the Ford-Wyoming involves a series of smaller problems that typically are solved internally. There are frequent underground water leaks, probably because of the land shifting from all of the clay being mined out. The building, which is more than sixty years old, needs regular maintenance. Replacement parts are expensive; for example, the iconic "Ford-Wyoming" lettering on the front of the main tower would take an estimated $1,000 per letter to replace.

There have been small fires, but nothing like the one in 1979. Everyday annoyances also took their toll. They had to outlaw barbecuing on site to avoid accidents with young children running around. Snowstorms mean you have to plow out at least half of the parking spaces for those drive-in fans who will attend no matter what the weather. Projectionists, walking up the steep industrial iron stairs, have dropped film reels, only to watch them unravel with alarming speed.

On the plus side, crime is limited. There have been burglary attempts, but nothing significant to note. Patrons occasionally get into fights, but those usually

The economy as well as the owner's decision to sell the property resulted in the closure of Screens 6 through 9 in 2010. As a result, the theater changed its name to the Ford Drive-in. *Waterwinterwonderland.com.*

Screens 6 through 9 came from other drive-in theaters in the area that the Shafer family owned at one time. *Waterwinterwonderland.com.*

are dispersed quickly by security or neighboring police. Shafer said that the only shooting he knew of happened when one moviegoer brought a raccoon with him to the show. Yes, a raccoon. "When that raccoon got loose, it scared the heck out of everybody. It got up in a tree, and it wouldn't come down. The guy told us it was his pet. I told him he shouldn't have brought it with kids in the area! I told him to go up in the tree and catch him. But we couldn't get it. Animal control couldn't get it. The police had to shoot it," Shafer recalled.

"You'd be amazed at how many pets people bring in. They bring in spiders. One guy bought in a snake; it was a twenty-foot boa constrictor. He got loose, too. The guy finally got him back in the car. That scared everybody to death. He told us the snake wouldn't hurt anyone," Shafer said, laughing. "When you deal with the public, you can expect anything."

People also really like to steal drive-in speakers. "We used to have people steal speakers every day of the week; it was especially bad on the weekends. They'd clip them off right at the base," Clark said. "Virgil and his staff on a Friday night would have to go through all of the theaters and replace them. It used to be a real hassle. It was so much so that during the late 1970s and 1980s, we were going and buying speakers from the closed drive-ins so we have spares." That is why you'll see speakers within the complex from drive-in theaters including the Gratiot, Bel-Aire and many more.

Occasionally, complaints about the theater's picture or quality also come up from time to time. Berean says that he mostly ignores those unless the person has a concrete issue—like the one time a patron said the bad guy was shot and died before his character was introduced. Berean said the movie had run for a week at this point when the discrepancy was finally noticed. "Every once in a while, we'll get complaints we are out of focus. Sometimes, they are right. Sometimes, they need to see an eye doctor," Berean said.

The biggest problem, if you can call it that, is exactly the same as the one that bothered the Clark family from the Ford-Wyoming's first night in business: people sneaking in to try to avoid paying the entrance fee for one or more of their group. "People come in, watch the movie and go home. We don't really have big problems. Well, the only problem we do have is people sneaking in. Anybody you talk to has snuck into a drive-in," said Berean, the theater's longtime manager during the Shafer era. "There are signs. For the most part, a teenager lady isn't coming by herself. So there's a guy in the trunk there. We've caught a lot of people that way…It's all fine and dandy. Keeps me busy."

Bill Clark says that they'll make a point to catch people from time to time, mostly to get the word out that sneaking in isn't tolerated. "We could

The Shafer family had to tear down the screens, projection booths and concession stand at its Wyoming location. *Waterwinterwonderland.com.*

make it a full-time job, chasing people around," he said. "One side of you says, 'They're stealing money from us and they didn't pay.' The other side of you says, 'We got the two or three who did pay.' If they feel good about the one that snuck in, then fine. You don't call the cops unless they get really belligerent. We're entitled to call, and they would come and respond. But why? You might alienate those two or three."

Getting away with sneaking into the Ford-Wyoming is practically impossible, Shafer notes. It's something that every staff members of the theater can recognize the minute they see it. "When we see a low-riding car, we watch," Shafer told the *Detroit Free Press* in 1985. "If they don't get out, they'll suffocate...People used to ask me if it bothered me when kids would sneak in by hiding in the trunks. Not really, because the first thing they'd do when they got out of the trunk was head to the concession stand."

CHAPTER 9

GOING DIGITAL

Well, parkin' in a drive-in movie
Wishin' life could be more groovy
I'd love to share my back seat, with a friend
Headlights flashin' in my mirror.
    —*Monkees, "Shake 'Em Up"*

Moviegoing, especially among drive-in theaters such as the Ford-Wyoming, has become a high-tech affair in recent years. Along with the innovations in sound that Charles Shafer and Bill Clark brought in on their own, the film industry's demand that all theaters convert to digital has been an expensive and challenging part of ownership.

Almost immediately after purchasing the Ford-Wyoming, Shafer and Clark began updating the theater with radio sound. The first phase involved burying cables under all of the theater's ramps, creating a giant antenna to share the sound with patrons through their AM radio. By the end of the decade, technology had advanced again, and now you could pull in the movie's sound through the clearer FM radio band. This was done through a simpler broadcasting antenna on top of each projection tower, Clark said. Every screen now has a sign that tells patrons which station to tune into. "You sound is as good as your radio. If you've got a Bose sound system, it will be crystal clear," Clark said.

Back then, there were holdouts who refused to use anything but the classic speakers. And there are a few left today, Clark noted. Some people didn't want to use their car radio because they feared their car batteries would die during the movie. If that does happen, Berean is handy with jumper cables.

"A lot of people rejected [radio broadcasts] for the first few years," Clark said. "We still need to keep speakers out now for the die-hards. There's a certain faction that won't use radio...Here we brought out this innovative technology, but they won't use it. I think it's nostalgia."

The biggest challenge came when the movie industry mandated that all theaters—both indoor and ozoners—convert to digital projection equipment. Hollywood imposed a drop-dead date of December 31, 2013, for drive-ins to convert from 35mm film. Converting to digital improved the picture, but it also cost about $75,000 per screen. In all, Bill Clark estimates the Ford-Wyoming paid between $350,000 to $400,000 to convert its five screens between 2012 and 2013.

"It was an expense we didn't need," Clark said, especially in light of how far attendance had fallen in recent years. "Had both Charles and I not had a lot of faith in the drive-in, we would have closed the doors." Drive-ins around the country closed because they couldn't afford to go digital, although it made sense for the movie industry, Clark noted. "They're shipping two steel cans with three small reels in each can all over the country, and most of it is airfreight. So, for example, a picture will get done Thursday night in California, and it opens Tuesday in Cleveland, so the freight bills are huge," Clark said. "Now they're a CD disc that comes in a heavily foamed package and maybe weighs a pound, pound and a half, instead of fifty to sixty pounds."

More than a decade ago, Hollywood created a kind of "rebate plan" with the savings from these shipping costs to help traditional indoor theaters with the digital transition. Drive-in theaters also receive this fee, worth somewhere around $700 each time a movie is shown, Clark said.

"Here's the problem," Clark said. "The interest we're paying on the $400,000 exceeds the rebates. What the film industry is doing is just paying the interest. We're paying the principal. At an annual rate, we might get $30,000. That sounds like a lot, but it's really not when you consider we're paying 6.0 to 6.5 percent interest on our loan [for the digital conversion]. It's not a profit center. The exhibitors like us took a hit."

Maintenance costs also went up—instead of only calling in help when things broke, now the machines have to be serviced on a regular basis, Clark said. And the drive-in's electric bill has gone up substantially with the switch to digital. "We used to use four-thousand-watt xenon bulbs; now we have to use six thousand. Logic tells you and I that six thousand watts takes more to operate than four thousand watts," Clark said. "The other downside to it

The conversion to digital meant the loss of film projectors, lovingly maintained by managers, including Ed Szurek at the Ford-Wyoming. *Kenny Karpov.*

The projection room was a breezy, loud space when it used film equipment before the digital conversion. *Kenny Karpov.*

is that the four-thousand-watt bulbs had a two-thousand-hour warranty on them. The six-thousand-watt bulbs only have a one-thousand-hour warranty. You're trading them out twice as fast, and they're burning more power."

Also, the digital conversion did not affect personnel costs, Clark said. You still need to have someone on site to make sure to press the button to play the movie on the jump-drive, which is part of a server the size of a refrigerator.

That said, Clark admits the change has been largely positive and that the Ford-Wyoming is the better for it now that it is done. "When drive-ins were first built, it was hard to see any scene that was dark or filmed at night. It really would get lost," Clark said. "Now you can see every scene because you're throwing more light out there. It's more like watching television during a dark scene…It's better for us because maybe now the theater-going public will come back to see a film with us more often."

The additional light helps because of how far a drive-in movie has to "throw the picture," so to speak, compared to indoor theaters, Shafer explained. In an indoor movie theater, the picture only has to go about 60 feet from the projector to the screen. At the Ford-Wyoming, the picture has to be thrown about 250 feet. Screen size also matters. An indoor screen might be 40 foot wide; the drive-in's screens are 80 feet wide, Shafer noted. And an indoor screen might be 12 feet high, whereas the Ford-Wyoming's are 40 feet high. "That [size] makes the light a lot less. Now, with digital, everything's even. They're the same indoor or outdoor because they're the same," Shafer said.

For Berean, the Ford-Wyoming's conversion was tinged with hopefulness. "When they put in these new projectors, I figured at least I have a little bit of a future. Maybe I'll be able to retire [from the Ford-Wyoming] now," Berean laughed. "The picture on the screen is 100 percent better. It's like watching your big-screen television at home."

Although the Ford-Wyoming had already converted, there were other theaters in Michigan that needed to make the change in the summer of 2013. One such conversion for the Cherry Bowl in Honor came through Honda Motor Company and its Project Drive-in. In all, Honda's fundraising program paid for ten theaters to convert to digital; another four were paid for through community donations, said Project Drive-in spokeswoman Alicia Jones.

Opposite, top: Film canisters were large, heavy beasts that sometimes showed up late to the Ford-Wyoming. *Kenny Karpov.*

Opposite, bottom: The projection booth shines another show, with the city's glow in the background. *Kenny Karpov.*

The ten drive-in theaters that received support from Honda's Project Drive-in were determined by more than 2 million votes at www.projectdrivein.com. Visitors to the site were encouraged to share Project Drive-in information with family and friends via social media, e-mail or texts and then also pledged to see one movie at their local drive-in and contribute to the national "save the drive-in" fund to help keep more drive-ins in business.

"We felt like we were in a unique situation to help because of the tie in between drive-ins and the car culture," said Jones, manager of social media at American Honda Motor Company Inc. "Our original idea was to help the one drive-in by our Ohio manufacturing plant, but we decided to expand the project when we realized it was an epidemic affecting all of these small businesses around the country."

Others have tried to step up and help as well. In 2014, songwriter Jimmy Buffett and the Coral Reefer Band performed live from the Coyote Drive-in in Fort Worth, Texas, and broadcast it live via satellite transmission to dozens of drive-in theater sites across the United States. This not only helped these drive-ins with ticket sales, but it also showed off their new digital equipment. The average participating drive-in was able to charge eighteen dollars per person, about double their normal prices.

Organizers said that technological advancements in digital projection and satellite delivery allowed these drive-in theaters to show live concert performances in high definition with the proper light output on outdoor screens. However, only about half of the nation's drive-ins could participate. There are 360 drive-in theaters left in the United States, and as of mid-2014, about 171 still had not made the transition from 35mm to a digital format, allowing them to compete with the conventional indoor theaters.

Going high tech resulted in the end of an era that many film historians and drive-in fans enjoyed. Gone are the film reels. There are no more giant film projectors. It is a change that is both welcome and strange for Berean. In fact, the Ford-Wyoming tried to give its projectors to the Henry Ford, a Dearborn museum known worldwide for its collection of automotive-related items.

Kristen Gallerneaux, curator of communication and information technology for the Henry Ford, visited Berean to collect what she could, including some speakers and the film reel that held the original concession advertisements, classics in their own right. When the Ford-Wyoming went digital, it had to buy a digital version of its dancing hot dog and juggling popcorn short, which is a beloved part of the movies in between shows.

With the new digital projector, Berean only has to swipe his finger around on the touch-screen to turn it on. Gallerneaux described it as

With radio frequency available for sound, the Ford-Wyoming's speakers are only used by die-hard fans. *Kenny Karpov.*

Speaker piles are kept within the storage areas of the drive-in for replacing those speakers that are missing or broken, despite having radio frequency. *Kenny Karpov.*

awakening "with a cold and even LED light—a docket of possibilities for film arrangements appear on the screen like entries on a recipe card. The screen acts as a digital slot system to swipe in and orchestrate trailers, advertisements and messages."

Berean uses something that looks like an external hard drive to load up everything he wants to show that night. "So you just take this thing, it sucks it up into the slot, extracts the files and away it goes! They told me I could make everything work from a laptop...I could just stay at home if I felt like it!" Berean said.

Gallerneaux said that she felt like Berean was conflicted about the ease of which the digital system worked. "Maybe it was a cool idea, operating it from home. But he likes being there. It's part of his everyday life. He's a part of that theater," she said. "The shift over to digital basically ended the profession of projectionist. That's definitely a lost art."

CHAPTER 10

"MY FAVORITE PLACE ON EARTH"

You're at a drive-in movie
With a cute brunette
A countin' on the kisses that you figure to get
Closer, closer, then she hollers ho!
Didja' ever get one of them girls
Who just wants to watch the show.
—*Elvis Presley, "Didja' Ever"*

The Ford-Wyoming serves as a meeting place for all of Metro Detroit. For families, it is a giant living room that offers kids a chance to play, watch a movie and then snooze comfortably in their car seats. For teens, the drive-in was their first coffeehouse, a place where they could meet without being accused of loitering. For young lovers, it offered privacy to share a movie, touch hands in the popcorn box and, perhaps, neck in the backseat.

Granted, much more than necking happens at the Ford-Wyoming. Many young lovers have consummated their relationships there. A few children may have been conceived. As Shafer says, you can do pretty much anything you want at a drive-in as long as you don't bother anyone.

People hold immense fondness for this Dearborn institution. Authors sing its praises in their books. Photographers of all stripes have images of it within their portfolios. Historians and architects revere its iconic main tower, marveling at its mix of Art Moderne and Art Deco styles. Preservationists relish that one of the remaining midwestern drive-ins has been kept so lovingly intact by its two owners.

Stop any random person on the street, and chances are they'll have a story about their time at this particular drive-in. You'll hear wistful tales of their youth, romantic stories about their dating histories and fond recollections about their childhood visits. And every story is told with this breathless awe as they recount the sights, sounds and smells of a night at the drive-in theater.

"I remember the first time I took my best friend, John, to the Ford-Wyoming. After we parked the car and made a walk to the concession stand, I won't forget his words: 'This is what I imagine heaven must be like,'" said Jason Millward, a former Detroit resident who has been going to the Ford-Wyoming since he was four years old.

"For a couple of movie buffs, he was absolutely right," Millward added. "On a cool summer night in the open air, everywhere you turn is another movie screen. Each showing a different film. If your only experience of watching flicks was of being in a giant dark room, the first time has got to be freeing."

Other countries struggle to understand America's love for the drive-in theater, said filmmaker and film historian Gary Rhodes, who teaches classes on the drive-in and others as a professor. "Where else but America would you go somewhere to just sit there? The functional purpose of a car is to drive it down the road. The purpose here is to drive and sit in it. It's such a strange idea to people in other countries. Of course, I herald it as an amazing thing," Rhodes said.

The love for drive-ins even goes as far as re-creating the sound, Rhodes noted. "There are even some films that I've seen released on DVD that have an alternative audio track that mimics a drive-in theater's speaker sound. They try to re-create that bad, mono, fuzzy sound. It's incredible," Rhodes laughed.

Drive-in theaters naturally attracted young couples and families in the post–World War II era, eager to enjoy their new vehicles and life on the road, said Al Binder, a senior editor for Ward's Automotive Group, a global authority on the latest news, data and analysis for the automotive industry. Binder said that some of his fondest childhood memories were of his brother sneaking kids into their local ozoner in the back of his trunk. "Up until World War II, people mostly traveled by train. They didn't go long distances; they stayed locally. But after the war, you see cars become more affordable. Suburbs start popping up. People started having more children—it was the baby boom. People just became more car orientated," Binder said.

The Ford-Wyoming has been the site of many photo shoots, including this one for clothing brand Spanish Fly. *Creative Direction: Eric Valdez. Photography: Doug Wojciechowski.* © *2014 Copyright Spanish Fly Detroit.*

"If you look back, you'll see all kinds of drive-in businesses. There are drive-in liquor stores, drive-in restaurants, drive-in churches. It was easier to go through a drive-in," Binder added. "When it came to movies, it made sense to throw everybody in the station wagon. You didn't have to get out of the car, and you didn't have to keep the kids quiet."

Detroit in particular gravitated to the drive-in culture, Binder said. "This was the Car Capital of the World. People here were orientated to getting in their cars and going somewhere. It's not like New York City, where people had subways. We never seemed to get mass transit right. The manufacturers were here, and everybody in Metro Detroit worked for them in one way or another. It was a major source of employment—still is. There's something about Detroit; cars are in our blood," Binder said.

Cars are in Metro Detroit's collective DNA, agreed Gary Ritzenthaler of Water Winter Wonderland. He says he'll go to a drive-in movie for the grandeur of the experience even if he doesn't necessarily like the movies showing. "We just love to have our personal vehicles. It's something about that autonomy, being under the stars and watching a movie on a giant

Royal Oak commercial photographer David Clement sought to catch the drive-in on a blue-sky day, when its colors pop. *David Clement.*

Photographer Kenny Karpov sought to create "timeless" photos of the drive-in during a photo shoot for *National Public Radio* station WDET. *Kenny Karpov.*

screen," Ritzenthaler said. "You can chat with people nearby, and the kids can play. You can't do that at indoor theaters. There, you're stuck inside, and you've got to be quiet."

Michigan's harsh winters also make its residents crave the outdoors and outdoor entertainment when the weather is warmer, said Johny Thomas, co-owner of the U.S.-23 Drive-in Theater in Flint Township. Thomas, thirty, and his partners took over the theater in 2009 after the death of longtime owner and founder Lou Warrington. "It's the freedom," he said. "In Michigan, you have to live as much as you can outside when it's warm here. Nothing against indoor theaters, but why would you want to be inside in the summer?…And there's nothing like a drive-in theater to take your mind off of everything that's going on in the world. You can just enjoy the night, hang with your kids, talk to your neighbors with the stars in the background."

The Ford-Wyoming was particularly memorable for April Wright, director of *Going Attractions: The Definitive Story of the American Drive-in*. Wright visited more than five hundred drive-ins and drive-in graveyards for her documentary, filming throughout much of 2007. Wright remembers being amazed at the long line of cars waiting to go into the Dearborn drive-in. "I had never been to Detroit at all. It was such an interesting city given its variety of classes and economic issues. We had driven through many of the city's neighborhoods when we arrived," Wright said. "I got to the Ford, and it was happening. It had huge, huge lines coming out of it at 3:00 a.m., and that was shocking. I hadn't seen drive-ins with that kind of line up to that point."

The Ford-Wyoming stood out then and now because of its size, Wright said. It still had its nine screens at the time, making it one of the world's largest drive-ins, with space for about three thousand cars. Plus, its original tower was immense for the time and especially these days, given the number of drive-ins that are now gone. "The tower is one of the best. It's just huge," Wright said. "It's one of the best original towers operating in the whole country, to be honest. There's nothing else like it—it's one of the most fascinating."

Author Sean Madigan Hoen included the Ford-Wyoming in several scenes of his 2014 book *Songs Only You Know: A Memoir*. The book, which received favorable reviews, is about his life as a young musician trying to hold his family together in the late 1990s Detroit area. The Ford-Wyoming appears as a refuge for the main characters:

> *Like everything else—the record stores and dollar movies and doughnut shops—I was waiting for the drive-in to close any day. There was a war on. A recession had begun, and you saw right away what it did to a town*

like Dearborn. The drive-in's all-night projections kept us company and soothed our fears. Or made us feel we were part of something, watching and waiting, straddling the edge of the city. "This," Scott said, "is my favorite place on earth."

The former Dearborn resident now calls Brooklyn home, but Hoen said that Ford-Wyoming looms large in his memories. "Ford-Wyoming is an essential piece of Americana, Metro Detroit–style. When you think of old drive-ins, you tend to think of lake towns and rural locales, as it seems that's why the surviving drive-in theaters have flourished," Hoen said. "The Ford-Wyoming, set there on the Dearborn/Detroit border, is a whole different experience."

He continued, "It's truly atmospheric; you can sense the seediness of downtown-bound Michigan Avenue to the southeast and the Rouge Steel factories and the abandoned storefront and a once-thriving, industrial city in which little capitalist glory remains. Drive-ins remind us of an era now passed, of car culture and hot dogs and first kisses—a time when humans had to leave the house to see film, any film, and what better way than to stare up at the giant actors as the moon shines down?"

He still visits whenever he comes to visit, driven to capture one more moment there. The Ford-Wyoming, he said, never disappoints. "As a kid in the '80s, the Ford-Wyoming spooked me a little because there were rumors of crime and violence and because no one had begun to cope with Detroit's dark age. It wasn't until later, as a young man, that I realized the magic of the place. It feels very 'of the people,' a cinematic car park in which the films can only attempt to contend with the unnamable enchants of the environment."

He added, "It's an experience, one you can only have there. It's never about the movies, for me. I'm absorbed in something else there, a feeling—a very Detroit feeling, at that. An intersection of what was and what is, and as though, were I to blink, the whole thing might disappear."

In an essay, author Debra M. Warrick of the Detroit Area Art Deco Society described the Ford-Wyoming as "both a physical and aesthetic symbol of America car culture at its best":

Created in the image of transportation, the modernistic style [Art Moderne and Art Deco] is often specific to those public buildings associated with the automobile. The smooth surfaces, curved corners and horizontal emphasis speak to all things Americans envision about their automobiles. More specifically, theater screen #1 and attached ticket booths are characterized by geometric angles, straight edges, shallow relief decorative elements and neon

The Ford-Wyoming's magic atmosphere had videographer Thomas Tomchak running for his camera during a family visit. *Thomas Tomchak.*

ornamentation. However, most impressive, is the screen's dramatic stepped façade emphasized by red and blue neon and porcelain enamel steel paneling.

There is a timeless feel to the Ford-Wyoming that draws artists and photographers. Kenny "Karpov" Corbin, a noted Detroit photographer, shot the drive-in for a *National Public Radio* story for the Detroit affiliate, and he spent three days on site. "The first time I drove by it, I couldn't believe how big it was, how large that tower was," Corbin said. "I loved everything about it. One of the best moments was walking up the stairs to the projection room, feeling the breeze on you and going through this door and hearing that projector. It was just beautiful to see that."

Because of its shift to digital, Corbin may have caught the final photographs of the Ford-Wyoming's original projectors. His goal during that assignment was to find a way to show the theater's eternal nature. "Everything about it is so authentic," Corbin said. "I was trying to take it from this era but make it look like it was from its original era. I wanted to show its agelessness through my lens. I wanted to transition it from today to an ageless drive-in and capture the history of it...They should preserve them. I know [some communities] aren't going to keep the old drive-ins, but I wish they would."

Photographer David Clement was drawn to the Ford-Wyoming on a sunny, blue-sky day. That is when the old drive-in tower looks its best, said Clement, who has provided photos for books such as *Detroit Art in Public Places*

Chad and Carla Whitton brought their four children to the Ford-Wyoming to introduce them to the drive-in tradition. *Carla Whitton.*

and his own words in *Talking Shops: Detroit Commercial Folk Art*, which highlights some of Detroit's most iconic retail signs.

"I'm interested in commercial archaeology. I'm interested in adaptive use of space and also in the history of drive-ins because basically they're almost all gone. I wanted to preserve them visually," Clement said. "Its

Modern façade against a blue sky—it just pops. So I had to stop to photograph it. It's much more like art that way…You have to walk around it, get that sense of space."

Videographer Thomas Tomchak of Wheaton, Illinois, did an impromptu photo shoot while visiting family in town one sunny evening. "As somebody who grew up in Dearborn, it felt like part of my own personal history. While driving by on the way back to my mom's house, I drove by it. Then, without giving it much though, I did a U-turn on Ford Road and pulled over. I grabbed my camera and just started taking photos. It was very impulsive. But the lighting was nice as the sun was nearly setting, and I felt like if I didn't do it right then I would probably never do it."

"Originally, I was just going to take photos from out front, but then I decided to try my luck and ask permission to go inside at the gate where you pay. They just waved me in, and I promised I would be quick (I don't think they really cared, to be honest). So, once inside, I started getting other photos as well so that I would have a good record of the place," Tomchak added. "The whole thing probably took twenty minutes. I think my wife thought I was nuts, but I was glad I did it when I processed the photos later. There is a kind of magic about that place, at least for me. I know it looks run-down and not well cared for, but it has a lot of history and character."

For Chris Heine, the Ford-Wyoming has always fascinated him because of its ornamentation and drama against a twilight sky. Heine spent more than a year creating his version of the Ford-Wyoming for the 2014 Modernism Exposition poster on behalf of the Detroit Area Art Deco Society. Heine does planning and architecture for the Smith Group inside Detroit's Guardian Building. "It's especially iconic when you're driving up to it at night. It's the colors, the lights. That's what I wanted to emphasize [in the poster]," Heine said. "It has a kind of a magic to it when you're pulling up to see a movie. It's not exotic, per se, but it's not something you see all of the time, so it grabs you. It's timeless because it's from a previous era, but it still shows the modern movies."

Todd Storrs is a "car guy"—he is an automotive car modeler for General Motors Corporation and restores classic vehicles both as an homage to his own love of cars and also to have something to drive on his regular visits to the Ford-Wyoming. His ride of choice most nights is a 1929 Model A hot rod. "I always get there early, get set up and then get some eats," Storrs said. "There's something very unique about that drive-in. It hasn't changed—you can go back to your old neighborhood, your old school, and it's all changed. It's been torn down, or it's not the same anymore. But the drive-in—it hasn't changed. There's that buzz in the air. There are the kids on the roof. People are there with their lawn chairs. It's exactly how I remember it."

Marilyn Gust Otten was a neighbor to the Clarks when they lived near her house on Meadowlane in Dearborn. Otten soon became the "official Clark babysitter, a title I was proud of," she said. "There were many summer evenings spent at the Ford-Wyoming with the Clarks," she added. "Mr. Clark, a hero in my eyes, provided treats from the concession stand, including my favorite—the foot-long hot dog! He also gave me posters of some of my favorite movie stars. Edd 'Kookie' Byrnes, of *77 Sunset Strip* fame, was my favorite. That poster was a treasure for a twelve-year-old girl!

"Later, during my teen years, the Clarks gave me many passes so my friends and I could spend summer evenings at the Ford-Wyoming. My dates were always impressed. 'You know the owners? Cool!'" Otten noted.

Katie and Brett Burghardt spent most of their courtship at the Ford-Wyoming, popping open the hood of his Jeep and watching movies like *Fever Pitch* and *Training Day*. They'd sit on the hood like kids, even wearing their pajamas on occasion, Katie said. "It was the classic date night with a double feature. Sometimes, we'd grab something from a restaurant and have dinner and a movie in the car," she added. "We'd watch the families with their lawn chairs, coolers and kids playing. We'd talk about how cool it would be to take our kids someday. Once we started doing it, we'd go all of the time."

Taking her four children to the drive-in theater was a landmark occasion for Carla Whitton. She remembers when her parents brought her and her four sisters to an outdoor movie; having such a large family means events like that were rare, indeed. "Back then, drive-ins were the only way we could afford to go to the movies. It was such a treat, and I wanted to have that experience for my kids," Whitton said.

Her kids did everything Whitton said she recalled doing on those special nights: ran around the open landscape, climbed into sleeping bags to cozy up in the car's trunk and laughed during the intermission between pictures— especially at the juggling popcorn and dancing soda pop, recalled her son, Jacob Whitton.

Millward, the longtime Ford-Wyoming fan and filmmaker, said that sharing the moviegoing experience with others is part of the drive-in. "Here's the thing about seeing films at the drive-in: it's an experience unto itself. In a regular theater, you go in and any distraction is a nuisance. The guy down the row who's crinkling the cellophane on his candy too loud during a quiet scene; kids having conversations with each other just behind you; the person next to you that is texting on their cellphone; loud crunchers or slurpers—if

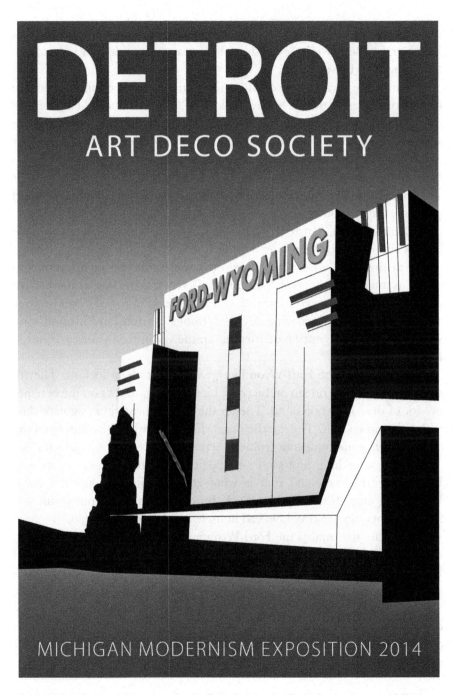

Chris Heine designed this poster for the Detroit Area Art Deco Society's annual Michigan Modernism Exposition. *Chris Heine.*

it pulls you out of the flick, you secretly hate them and it ruins your day. However, at the drive-in, everything is texture and adds positively to the atmosphere," Millward said.

"For instance, a rainstorm during a film is the best—and this goes double for horror movies in the fall before Halloween," he continued. "The noise of the rain on the roof your car; the sound of the windshield wipers; the movie screen distorted by the rain running down the windshield suddenly clear for a split-second and the wipers fly by—that enhances the experience. Even on a clear day, you're likely to get a flock of seagulls flying by. You'll hear their squawking, and then they'll fly between the projector and the screen, and you'll see their silhouettes going by briefly."

Millward added, "Instead of being annoying, it just brings you back to how fun it is to be watching a giant film in front of you in the dark of night, in the open air. Even obnoxious people who are talking nonstop aren't as much of a worry. It's all atmosphere. And the louder people are around you, the more comfortable you feel speaking to your pals or your girlfriend without whispering. Even if things get too loud around you, you always have the option of turning up the volume on your car's stereo to drown them out."

So what makes the Ford-Wyoming great? It's all about its fans. They're the ones who would ride in on bicycles, hit the swings and sit on the benches outside of the concession stand. They're the ones in the station wagon packed with kids and cousins. They're the ones who would watch the film from the back of their grandma's convertible or scream at *Terminator 3* in a Delta '88.

There is something profoundly nostalgic about the drive-in experience, Ritzenthaler believes, and that is what keeps institutions like the Ford-Wyoming going. "It's like a throwback. It takes you back to your youth and the days when the country believed in itself a little more," he said.

It's the mix that makes the Ford-Wyoming. It's the families who need an inexpensive night out. It's the hipsters who like the rust and peeling paint. It's the teens who still need a place to go out on a date. It's the tried-and-true fan who wants the drive-in to stay, no matter what the cost. Said Storrs, "When I go there, I always go to the snack bar and buy too much. I'll buy more popcorn than I can eat. Because if I buy more, they can afford to stay open another season. If I don't do that, then the next time I want to go, it won't be there."

A COMMUNAL EXPERIENCE

Drive-in movies, Friday nights
Drinkin' beer and laughin'
Somehow things were always right.
   —*Jim Croce, "Alabama Rain"*

W hat is the future of Dearborn's iconic drive-in theater? What will
   happen to the Ford-Wyoming when Charles Shafer, who is now in his
nineties, decides to retire? What will future generations do without the big
screens or neon marquees to light the way?

Luckily for Metro Detroit, the Ford-Wyoming has several things going for
it: It is located in an industrial zone with property unsuitable for pretty much
every kind of development. It has regular, though smaller, crowds who adore it.
And most importantly, it has owners who never want to see it close. "It gives me
something to do. I wouldn't want to sit around and knit sweaters. I love what I'm
doing," Shafer said. "My whole theory [about this business] was no kid wants to
take a date and see a movie at their mother and dad's house. There always has to
be a drive-in…I hope this never closes; as long as people come, it will stay open."

His longtime business partner agreed. "We have faith in it," said Bill Clark.
"There's still a lot of life in the old girl…We've had offers to sell it over the
years, but they were never those staggering numbers we had to think about.
They might be for a couple million dollars. But Charles and I agree: if we
keep it five more years, we will make that much anyway. So why sell it?…We
just got to keep the people coming back."

The main tower continues to serve as an inspiration to current and future generations. *Kenny Karpov.*

Manager Virgil Berean says that he hopes to retire from the Ford-Wyoming, and owners Charles Shafer and Bill Clark are likely to fulfill his wish. *Kenny Karpov.*

That's it in a nutshell, isn't it? We've just got to keep going back. "We get lazy about going to the theater, and we're missing part of that crucial communal experience," said Kristen Gallerneaux of the Henry Ford. Yes, they are still building drive-ins. But they are mostly inflatable screens that someone throws up in a cornfield, with a hot-dog stand out front. Sure, that's an exaggeration, but you get the point. Buildings like the Ford-Wyoming are expensive to build. They were born out of a bygone era when going to the movies meant something. The experience started at the sidewalk, and it continued through to the field where you'd park to see the movie. It was magic. It *is* magic.

"If they ever stake down that big concrete screen, nobody's going to build one like that again," said Robert Thompson, director of the Bleier Center for Television & Popular Culture at Syracuse University. The Ford-Wyoming is one of a kind, he noted. "When the drive-in movie concept developed, one had to think, 'There is no possible way to come up with a better marriage,'" Thompson said. "It's the two things that Americans love most: their cars and their movies. On top of that, you get to eat junk food. It seemed like nothing could trump it."

Things have tried. Ephemeral things like *Laverne & Shirley*, iPhones, "Candy Crush" and the World Wide Web. Those things might entertain you for a moment. A drive-in movie and its memory of a warm night, your mother's arms and a funny movie will last forever, evoking feelings of love, youth and togetherness. Pretty hard to top that.

"I've never been to Disneyland. While I'm sure I'd enjoy it, I'm not sure if it would be time better spent than at the Ford-Wyoming. To me, it is the happiest place on earth," said superfan Jason Millward. "It's an outdoor multiplex that turns every film into a genuine experience. And on those rare occasions where you can get a horror double-feature, and you bring your significant other…well, it's like going back in time to the golden age of drive-in theaters in their hey-day of low-budget genre films. Dearborn truly is home to a national treasure."

The Ford-Wyoming is probably open tonight. Why don't you go?

OPEN MICHIGAN DRIVE-IN THEATERS

CAPRI DRIVE-IN
119 West Chicago Road, Coldwater, MI, 49036

CHERRY BOWL DRIVE-IN
9812 Honor Highway, Honor, MI, 49640

5-MILE DRIVE-IN
28190 State Route 152, #M Dowagiac, MI, 49047

FORD DRIVE-IN
10400 Ford Road, Dearborn, MI, 48126

GETTY DRIVE-IN THEATRE
920 East Summit Avenue, Muskegon, MI, 49444-3206

HI-WAY DRIVE-IN
2778 East Sanilac Road, Carsonville, MI, 48471

SUNSET AUTO THEATER
69017 Red Arrow Highway, Hartford, MI, 49057

U.S.-23 DRIVE-IN
G-5200 Fenton Road (Old US-23), Flint, MI, 48507

SELECTED BIBLIOGRAPHY

Burton, Clarence Monroe, Gordon K. Miller and William Stocking, eds. *The City of Detroit, Michigan, 1701–1922*. Vol. 2. Detroit, MI: Forgotten Books, 1922

———. *The City of Detroit, Michigan, 1701–1922*. Vol. 3. Detroit, MI: Forgotten Books, 1922. .

Cahill, Marie. *A History of Ford*. N.p.: Brompton Books Corporation, 1992.

Danziger, Sheldon, Reynolds Farley and Harry J. Holzer, eds. *Detroit Divided*. New York. Russell Sage Foundation Publications, 2002.

Davis, Michael W.R. *Detroit's Wartime Industry: Arsenal of Democracy*. Charleston, SC: Arcadia Publishing, 2007.

Freund, David M.P. *Colored Property: State Policy and White Racial Politics in Suburban America*. Chicago: University of Chicago Press, 2007.

Gavriolovich, Peter, and Bill McGraw, eds. *The Detroit Almanac: 300 Years of Life in the Motor City*. Detroit, MI: Detroit Free Press, 2006.

Good, David L. *Orvie: The Dictator of Dearborn—The Rise and Reign of Orville L. Hubbard*. Detroit, MI: Wayne State University Press, 1989.

Haigh, Henry A. *Early Days in Dearborn*. N.p., 1920.

Hauser, Michael, and Marianne Weldon. *Detroit's Downtown Movie Palaces.* Charleston, SC: Arcadia Publishing, 2006.

Hoen, Sean Madigan. *Songs Only You Know: A Memoir.* New York: Soho Press. 2014.

Hutchison, Craig, and Kimberly Rising. *Dearborn, Michigan.* Charleston, SC: Arcadia Publishing, 2003.

Kenyon, Amy Maria. *Dreaming Suburbia: Detroit and the Production of Postwar Space and Culture.* Detroit, MI: Wayne State University Press, 2004.

Leake, Paul. *History of Detroit: A Chronicle of Its Progress, Its Industries, Its Institutions and the People of the Fair City of the Straits.* N.p.: Lewis Publishing Company, 1912.

Lewis, David W. *Eddie Rickenbacker: An American Hero in the Twentieth Century.* Baltimore, MD: Johns Hopkins University Press. 2008.

Metress, Seamus P., and Eileen K. Metress. *Irish in Michigan.* East Lansing: Michigan State University Press, 2006.

The Michigan Alumnus 61, no. 15 (March 12, 1955).

Poremba, David Lee. *Michigan: On the Road Histories.* N.p.: Interlink Books, 2006.

Segrave, Kerry. *Drive-In Theaters: A History from Their Inception in 1933.* Jefferson, NC: McFarland & Company, 1992.

Successful Men of Michigan: A Compilation of Useful Biographical Sketches of Prominent Men. N.p.: Collins, 1914.

Time-Life Books, eds. *The American Dream: The 1950s.* N.p.: Time Life Education, 1997.

Young, Coleman, and Lonnie Wheeler. *Hard Stuff: The Autobiography of Mayor Coleman Young.* N.p.: Viking Adult, 1994.

INDEX

S

T

W

Y

ABOUT THE AUTHOR

Karen Dybis is a Metro Detroit writer who has blogged for *Time* magazine, worked the business desk at the *Detroit News* and jumped on breaking stories for publications including *Corp! Magazine* and *Detroit Unspun* and the *Agence France-Presse* newswire. She was born in Bad Axe (where she saw her first drive-in movie in the back of a Suburban), raised in Romeo, worked on Mackinac Island and hangs out in St. Clair Shores. And she is proud to say that her two children had their first drive-in movie experience at the Ford-Wyoming.

CPSIA information can be obtained
at www.ICGtesting.com
Printed in the USA
LVOW04*1519191017
553027LV00030B/1057/P